INTEGRATED REGIONAL MODELS

INTEGRATED REGIONAL MODELS

INTERACTIONS

BETWEEN

HUMANS

AND THEIR

ENVIRONMENT

EDITED BY

PETER M. GROFFMAN
GENE E. LIKENS

CHAPMAN & HALL
ITP An International Thomson Publishing Company

New York • Albany • Bonn • Boston • Cincinnati
• Detroit • London • Madrid • Melbourne • Mexico City
• Pacific Grove • Paris • San Francisco • Singapore
• Tokyo • Toronto • Washington

First published by
Chapman & Hall
One Penn Plaza
New York, NY 10119

Published in Great Britain by
Chapman & Hall
2-6 Boundary Row
London SE1 8HN

Printed in the United States of America

Library of Congress Cataloging-in-Publication Data

Integrated regional models : interactions between humans and their
   environment / editors Peter M. Groffman, Gene E. Likens.
      p. cm.
   Includes bibliographical references.
   ISBN 0-412-04811-6
      1. Environmental policy—Mathematical models.   2. Human ecology—Mathemat-
ical models.   3. United States—Climate—Case studies.   4. Regionalism—Mathematical models.   I. Groffman, Peter M., 1958–
   II. Likens, Gene E., 1935– .
   GE170.I58 1994
   363.7'001'5118—dc20   93-44573
                                                                                    CIP

British Library Cataloguing in Publication Data available

Please send your order for this or any **Chapman & Hall book to Chapman & Hall, 29 West 35th Street, New York, NY 10001, Attn: Customer Service Department.** You may also call our Order Department at 1-212-244-3336 or fax your purchase order to 1-800-248-4724.

For a complete listing of Chapman & Hall's titles, send your requests to **Chapman & Hall, Dept. BC, One Penn Plaza, New York, NY 10119.**

Contents

Preface

This book presents results from an international workshop on "Integrated regional models: analysis of interactions between humans and their environment" that was held on 4–8 October 1992 at the Institute of Ecosystem Studies in Millbrook, NY. The workshop brought together 38 scientists from biological, physical and social sciences in roughly equal numbers to assess the role of integrated regional models (IRMs) for dealing with complex, environmental problems. Integrated regional models are conceptual and mathematically based models that include, within the structure of the model, mathematical descriptions of the physical environment, biological interactions and human decision-making and its consequences. Funding for the workshop was provided by the National Science Foundation.

An overall goal of the workshop was to promote communication and research interactions among scientists in these three disciplines. This was the first workshop that many had attended where there was more than a token number of social scientists at a meeting of ecologists, a token number of ecologists at a meeting of physical scientists, etc. There was an unusual sense of excitement and interchange at this workshop.

The specific workshop objectives were:

1. To inform scientists from the different disciplines about the current status of knowledge of processes operating at a variety of scales, including regions.
2. To explore how scientists from different disciplines can work together to advance their collective knowledge of critical processes and to improve models within and among their disciplines.
3. To identify the steps needed to move toward the development of integrated regional models that represent linkages among biological, physical and human systems.

4. To identify the data requirements necessary to do successful integrated regional models.

The workshop format was one of plenary talks and small group discussions. Plenary speakers were paired as an "outsider" to models in the discipline being considered and an "insider," who presented a case study from within the discipline. All participants in the workshop had been sent a reading list of representative background papers prior to the workshop (see Table 1). These publications provided state-of-the-art examples and perspectives characterizing the current status of published models for each of the three disciplines. This approach was used to foster communication and understanding and worked well.

The plenary talks, one of the conference workgroup discussions, and the conference summary talk are contained in this book. The first group of papers contains an overview of biological models from the perspective of a physical scientist, an ecologist's view of some social science models, and a social scientist's perspective of physical science models. The second group of papers consists of regional scale case studies in social, ecological and physical science and a discussion of the prospects for the development of integrated regional models for forested regions. This latter paper arose from a conference "workgroup" discussion.

Because management decisions frequently are based on piecemeal information from a disciplinary approach, there is a real need to gain understanding at larger, integrated spatial scales. Thus, IRMs may indicate approaches both to integrated basic science (conceptual, experimental, mathematical, simulation) on a regional scale and to integrated regional *management*. Currently, however, there is a significant intellectual void relative to development and application of IRMs, which raises two important questions. Is it best to integrate among disciplines at the beginning or at the end of the modeling process? Are IRMs better done for region-specific or problem-specific questions? Also, the terms interdisciplinary and multidisciplinary are used loosely among the various disciplines. Complex problems are usually viewed from the vantage point of specific disciplines rather than at the interstices between disciplines. Scientists rapidly retreat to the "safety" and familiarity of their own disciplines when faced with problems of extreme complexity. Thus, interdisciplinary efforts remain the goal of integrated approaches, but multidisciplinary efforts are the current status. These and other questions and issues were wrestled with by the participants at the workshop.

The workshop participants believed that the outcome of a successful IRM activity would be a substantial advancement in the collective ability to adapt to a rapidly changing world. There would be a better appreciation of other disciplinary approaches and concerns that are contributing to regional issues. There would be an opening of dialogue among groups that previously were largely isolated. This interaction would result in the development of active cooperation among the disciplines and stimulate thinking far beyond current boundaries. As a result, IRMs would create unusual opportunities for the cross fertilization of disciplinary ideas and concepts.

Table 1. Reading List for Integrated Regional Models Workshop

Biological Discipline

Bonan, G. B. 1991. Atmosphere-biosphere exchange of carbon dioxide in boreal forests. *Journal of Geophysical Research* **96**(D4):7301–7312.

Burke, I. C., C. M. Yonker, W. J. Parton, C. V. Cole, K. Flach, and D. S. Schimel. 1989. Texture, climate and cultivation effects on soil organic matter content in U.S. grassland soils. *Soil Science Society of America Journal* **53**:800–805.

Costanza, R., F. H. Sklar, and M. L. White. 1990. Modeling coastal landscape dynamics. *BioScience* **40**:91–107.

Greenwood, D. J., and A. Walker. 1990. Modelling soil productivity and pollution. *Phil. Trans. R. Soc. Lond. B* **329**:309–320.

Parton, W. J., D. S. Schimel, C. V. Cole, and D. S. Ojima. 1987. Analysis of factors controlling soil organic matter levels in Great Plains grasslands, USA. *Soil Science Society of America Journal* **51**:1173–1179.

Pastor, J., and W. M. Post. 1988. Response of northern forests to carbon dioxide induced climate change. *Nature* **334**:55–58.

Rizzo, B., and E. Wiken. 1992. Assessing the sensitivity of Canada's ecosystems to climatic change. *Climatic Change* **21**:37–55.

Running, S. W., R. R. Nemani, D. L. Peterson, L. E. Band, D. F. Potts, L. L. Pierce, and M. A. Spanner. 1989. Mapping regional forest evapotranspiration and photosynthesis by coupling satellite data with ecosystem simulation. *Ecology* **70**:1090–1101.

Schimel, D. S., W. J. Parton, T. G. F. Kittel, D. S. Ojima, and C. V. Cole. 1990. Grassland biogeochemistry: links to atmospheric processes. *Climatic Change* **17**:13–26.

Schimel, D. S., T. G. F. Kittel, and W. J. Parton. 1991. Terrestrial biogeochemical cycles: global interactions with the atmosphere and hydrology. *Tellus Series AB* **43**:88–203.

Tucker, C. J., I. Y. Fung, C. D. Keeling, and R. H. Gammon. 1986. Relationship between atmospheric carbon dioxide variations and a satellite-derived vegetation index. *Nature* **319**:195–199.

Urban, D. L., R. V. O'Neill, and H. H. Shugart. 1987. Landscape ecology: A hierarchical perspective can help scientists understand spatial patterns. *BioScience* **37**:119–127.

Physical Discipline

Dickinson, R. E., R. M. Errico, F. Giorgi, and G. T. Bates. 1989. A regional climate model for the western United States. *Climate Change* **15**:383–422.

Giorgi, F., and L. O. Mearns. 1991. Approaches to the simulation of regional climate change—a review. *Reviews of Geophysics* **29**:191–216.

Jacob, D. J., and S. C. Wofsy. 1988. Photochemistry of biogenic emissions over the Amazon forest. *Journal of Geophysical Research* **93**:1477–1486.

Pielke, R. A., G. A. Dalu, J. S. Snook, T. J. Lee, and T. G. F. Kittel. 1991. Nonlinear influence of mesoscale land use on weather and climate. *Journal of Climate* **4**:1052–1069.

Shukla, J., C. Nobre, and P. Sellers. 1990. Amazon deforestation and climate change. *Science* **247**:1322–1325.

Trenberth, K. E., G. W. Branstator, and A. Arkin-Phillip. 1988. Origins of the 1988 North American drought. *Science* **242**:1640.

Vorosmarty, C. J., B. Moore, A. L. Grace, M. P. Gildea, J. M. Melillo, B. J. Peterson, E. B. Rastetter, and P. A. Steudler. 1989. Continental scale models of water balance and fluvial transport an application to South America. *Global Biogeochemical Cycles* **3**:214–266.

continued

Table 1   Continued.

Social Discipline

Bostrom, A., B. Fischoff, and M. G. Morgan. 1991. Characterizing mental models of hazardous processes: A methodology and application to radon. Prepublication copy. *Journal of Social Issues* **48**:85–100.

Edmonds, J., and J. Reilly. 1983. Global energy and CO_2 to the year 2050. *The Energy Journal* **4**:21–47.

Lee, R. G., R. Flamm, M. G. Turner, C. Bledsoe, P. Chandler, C. DeFerrari, R. Gottfried, R. J. Naiman, N. Schumaker, and D. Wear. 1992. Integrating sustainable development and environmental vitality: A landscape ecology approach. *In* R. J. Naiman (ed.). *Watershed Management: Balancing Sustainability and Environmental Change.* Springer-Verlag, New York, pp. 497–518.

Grossman, W. D. 1991. Model- and strategy-driven geographical maps for ecological research and management. *In* P. G. Risser and J. Mellilo (eds.). *Long Term Ecological Research: An International Perspective, Scope 47.* John Wiley & Sons, New York, pp. 241–256.

Marknsen, A. 1986. Where and why high tech locates. *In* A. Galsmeier, P. Hall, and A. Marknsen (eds.). *High Tech America.* Allen & Unwin, Boston, pp. 144–169.

Nordhaus, W. D. 1991. To slow or not to slow: The economics of the greenhouse effect. *The Economic Journal* **101**:920–937.

Nordhaus, W. D., and G. Yohe. 1983. Future carbon dioxide emissions from fossil fuels. *In Changing Climate: Report of the Carbon Dioxide Assessment Committee.* National Academy Press, Washington, pp. 87–153.

Portney, P., and A. Krupnick. 1991. Controlling air pollution: A benefit-cost assessment. *Science* **252**:522–528.

Reilly, J., J. Edmonds, R. Gardner, and A. Brenkert. 1987. Monte Carlo analysis of the IEA/ORAU energy/carbon emissions model. *The Energy Journal* **8**:1–29.

Southworth, F., V. Dale, and R. V. O'Neill. 1991. Contrasting patterns of land use in Rondonia, Brazil: Simulating the effects on carbon release. *International Social Sciences Journal* **130**:681–698.

To affect this level of activity there are a number of fundamental needs to be considered. These needs relate to the availability of information (e.g., data) in a usable form, availability of knowledge and expertise, the ability to develop a common language for effective communication (e.g., overcoming jargon), and the identification of a common currency for model validation. Finally, there is a real need to develop several successful activities (e.g., case studies) for the broader community to learn from and to build upon.

We hope that this book will serve as a first step and motivating force for the development of IRMs. The cross-disciplinary reviews and case studies should go far to bridge the intellectual and practical gaps between ecological, physical and social scientists by providing concrete examples of how scientists in different disciplines approach common problems. We hope that the development of IRMs will produce advancements in environmental sciences and help in the solution of contemporary problems relating to human interactions with the environment.

PETER M. GROFFMAN
GENE E. LIKENS

Contributors

Richard Berk
Department of Sociology
and Interdivisional Program in Statistics
University of California
Los Angeles, CA 90024

Ingrid C. Burke
Natural Resources Ecology Laboratory
Colorado State University
Ft. Collins, CO 80523

William L. Chameides
School of Earth and Atmospheric Sciences
Georgia Institute of Technology
Atlanta, GA 30332

C. Vernon Cole
Natural Resources Ecology Laboratory
Colorado State University
Ft. Collins, CO 80523

Peter M. Groffman
Institute of Ecosystem Studies
Box AB
Millbrook, NY 12545

Bruce Hayden
Department of Environmental Sciences
Clark Hall
University of Virginia
Charlottesville, VA 22903

William K. Lauenroth
Natural Resources Ecology Laboratory
Colorado State University
Ft. Collins, CO 80523

Gene E. Likens
Box AB
Institute of Ecosystem Studies
Millbrook, NY 12545

Diana Liverman
Department of Geography
Pennsylvania State University
University Park, PA 16802

Stephen W. Pacala
Department of Ecology and Evolutionary
Biology
Princeton University
Princeton, NJ 08544

William J. Parton
Natural Resources Ecology Laboratory
Colorado State University
Ft. Collins, CO 80523

David S. Schimel
National Center for Atmospheric Research
P.O. Box 3000
Boulder, CO 80303

I

Introduction

1

Introduction
David Schimel

Integrated regional models (IRM) are conceptual and mathematical models that may include, depending upon the application, components that describe the physical environment, biological interactions and human decision-making and its consequences. Disciplinary, regional or partly coupled models now exist in the atmospheric, ecological and social sciences. Work is in progress to couple these disciplinary models in response to a variety of contemporary environmental problems. Progress on coupling has been limited by lack of interaction among disciplines, lack of an accepted interdisciplinary framework, differences in technical approaches, and the absence of appropriate supporting data sets. In many cases, coupling is also inhibited by inadequacies of the disciplinary science, in that, in many cases, disciplinary studies have not addressed the processes that couple the system of focus to other system components. In many models, the coupling variables are not simulated as well as the internal dynamics within the component (Parton et al., in press). The severity and complexity of regional environmental problems, and the significant contribution of regional problems to global environmental issues dictates that progress in interdisciplinary and collaborative modeling be accelerated.

Despite concerns about "global change," many key scientific and policy issues related to the environment are developed, considered and ultimately implemented on regional scales—local jurisdictions, watersheds, agricultural production regions. At these levels, regional interactions of physical, biological and social systems play a critical role in defining the problems that must be resolved. Interactions at the regional scale provide the context for much political decision-making on environmental issues.

There is a great need for information about coupled systems in political decision-making. This process is dominated currently by concerns over single issues, which in many cases are merely components of complex, linked problems. Environmental issues that require an integrated analysis include coastal zone management, regional atmospheric chemistry and pollution, non-point-source pollution, commodity production in forested landscapes, and urban growth.

While most environmental problems are rooted in physical or biological science, e.g., the production of pollutants, the harvesting of commodities, degradation of biological resources, changes in temperature or precipitation, they are driven by human behavior. It is not feasible to determine how problems arose or how they can be solved without understanding the human decision-making process. Physical or biological "fixes" can be prescribed for environmental problems, but implementation of these fixes at regional or global scales depends upon human behavior. Typically, environmental management requires making trade-offs, and deciding upon appropriate trade-offs requires both agreed-upon value systems and credible estimates of the true environmental effects of decisions.

The nature and the extent of coupling between disciplines in IRM are varied and complex. Interactions between models can range from simple use of the output from one type of model in another model, to complex feedbacks between families of models. For example, while social scientists can provide relatively straightforward information on adoption of a specific remediation practice as input to physical or biological models, analyzing feedbacks related to the practice is more complex. Developing models capable of depicting such complex feedbacks requires a high degree of coordination among scientists from different disciplines, as well as efforts by scientists versed in several disciplines.

There are many impediments to the development of IRM capable of addressing contemporary environmental problems. The regulatory agencies that are responsible for dealing with environmental problems may lack the scientific scope in terms of expertise, disciplinary training and number of staff to undertake a major development effort for IRM. Development should take place within a broad academic arena including agency and academic researchers, managers, and policy makers. A major objective of this book is to foster such interactions and generally "set the wheels in motion" for accelerated interdisciplinary IRM development.

Within the academic community there are major obstacles to interdisciplinary interaction. University departments continue to be organized along traditional disciplinary lines rather than on specific problems (Hall, 1992), e.g., there are no departments of "non-point-source pollution." Similar to academic institutions, funding agencies are often organized along traditional disciplinary lines that inhibit the funding of large, multidisciplinary projects. The emergence of "global change" research centers at many universities may be a positive step toward the development of multidisciplinary, problem-oriented research groups. An objective of this book is to establish a structure for IRM development that can serve as a nucleus for the organization of these groups.

This chapter will introduce several issues in regional modeling of atmospheric and ecological systems, specifically, the large-scale context of regional systems, scaling of atmospheric processes to regional scales, sensitivity and importance of coupling parameters, and data needs for regional analysis. I will not address models of economics or human decision-making, but will rather focus on the natural science components of regional models.

Large-Scale Context of Regional Systems

In both atmospheric science and ecology, regional models (mesoscale and ecosystem models in disciplinary terminology) have focused on systems where internal interactions are much more important than external in controlling system dynamics (Schimel et al., 1991). This assumption, while convenient, is invalid in coupled or integrated regional models and is a major impediment to progress.

For example, in the atmosphere, regional-scale processes evolve over time as a function of changing influences from the global atmosphere. The global influence of the El Niño phenomenon, which arises from processes in the southern ocean, but affects climate worldwide on an episodic basis, is an excellent example of such connectivity. The origins of the 1988 drought in the mid-continental United States was likewise caused by remote influences in the atmosphere–ocean system, again including effects from the tropical oceans (Trenberth et al., 1988).

Ecological processes are influenced by remote influences. Inputs of nutrients transported through the atmosphere are significant influences over the behavior of both perturbed and relatively pristine systems (Aber et al., 1989; Langford et al., 1992). Changes in atmospheric concentrations of CO_2, NH_4, SO_4 and other compounds are influenced by regional, physical and biological processes, and influence ecosystems worldwide, through fertilization, toxicity and climate. Movement of organisms can also influence the dynamics of ecosystems by altering structural and functional properties of ecosystems. Examples where changes in the distributions of organisms influence ecosystem processes include the invasion of grassland ecosystems by woody plants as a consequence largely of grazing and the influences of the N-fixing exotic *Myrica faya* in Hawaii on system-level N cycling (Archer et al., 1988; Vitousek 1990). In both cases (shrub invasion and *Myrica*), the dynamics of specific ecosystems were influenced by the mobility of vegetation at quite large scales.

When considering ecosystem dynamics at large spatial scales (e.g., regional) and over long periods of time (decades to centuries), it is difficult to understand regional processes without considering the larger-scale context, and how that may change over time. When considering both atmospheric and ecological processes over periods of time associated with climate and climate change (10–100 years), it is impossible to understand or forecast regional processes without considerable understanding of their global context. IRMs are not a substitute for improved global models of energy, biogeochemistry, economics and decision making but rather are "magnifying lenses" for the examination in detail of specific areas.

Scaling of Atmospheric Processes

Projections of global climate are made using models with spatial resolutions of 200–500 km. These models are not very successful at predicting weather as people experience it or at the scales that measurements are made. Rather, these models stimulate the large-scale features of the climate system, the general circulation, major storm tracks and meridional gradients.

In order to use output from atmosphere general circulation models (GCM) ("climate models") in regional models of hydrology, ecology and agriculture, the finer scale features of climate must be inferred. Several techniques exist. These include the use of "model output statistics" (MOS) in which observed large-scale climate features are related to observed local climate. The same correlation structure is then assumed to apply between GCM-simulated large-scale features and local climates. Others have used "weather generators," which include effects of topography and elevation to modify coarse climate descriptions for specific sites (e.g., Bretherton et al., 1992; Running et al., 1987). Neither of these techniques is fully satisfactory, since in the first technique the assumption of global–local relationships is required, and in the second technique effects larger than the site and smaller than the global data set (such as rain-shadows) are not accounted for.

Recently, a third technique has been advanced, in which a higher-resolution model (1–100 km resolution) is forced at its boundaries by either data or model results from the global scale. The higher-resolution model can stimulate the effects of clouds and small convective systems, topography, vegetation and land–water contrasts with large-scale features of climate constrained by the boundary conditions. This technique—referred to as nesting—avoids some of the problems that affect MOS and weather generators. Nesting preserves more of the physical interactions, and the nested model may include a more-detailed representation of topography and land surface properties than a global model, allowing for better simulation of feedbacks (Georgi, 1990). Nested simulations often improve simulation of rainfall, especially as it is affected by orography. It is not often appreciated how coarse the representation of topography is in GCMs. For example, in most GCMs the western United States is represented as a single broad ridge. Georgi (1990) in his nested model used a 60 km resolution and was able to resolve the large-scale structure of the Sierras, the Rockies and some of the major Basin and Range mountain chains.

In the nesting procedure, the nested model usually uses essentially the same representation of physics (for consistency) that the global model does, but includes greater detail of orography, vegetation and surface hydrology. When details of regional climate are modeled poorly in a global model because of lack of detail in surface properties, the nesting approach can improve simulation of climate significantly. However, when regional climates are poorly simulated because the global model is faulty, they merely provide detailed but fallacious regional climates. For example, regional variations in Sahelian and East Asian Monsoon regions are poorly represented in global climate models because of many faults in the global models, including inadequate simulation of the intertropical convergence zone, air–sea–land temperature contrasts governing the monsoon and other factors not amenable to improvement via nesting. Application of nesting to climate simulations has so far been a one-way procedure, in which data are fed into the high-resolution model at its boundaries with no feedback. It is possible that phenomena that occur intrinsically at the regional scale have consequences for global climate. For example, circulations induced by orography, land–sea or vegetation contrasts may influence global circulation patterns. One-way nesting cannot capture such effects, and, although in principle two-way nesting is possible, it remains a possibility for the future.

In summary, the problem of developing credible regional forecasts of climate remains an unsolved problem (Houghton et al., 1990). However, techniques are being developed to produce improved regional scenarios from global models that are applicable in at least some regions. While this process is frustrating in its difficulty, it has resulted in exciting dialogue. Ecologists have explained to climatologists how they use climate data and what their requirements are, and climatologists have presented them with a suite of helpful techniques (Bretherton et al., 1992). While the availability of regional climate results for ecological, hydrological and agricultural modeling is far from satisfactory, the situation is much improved over a few years ago. The mismatch between climate results and ecological requirements has produced some exciting research.

Sensitivity to Coupling Parameters

Integrated regional models are composed of some combination of coupled submodels of biological, physical, chemical and human systems. These submodels, usually developed more or less in isolation, must be modified to communicate key information among themselves. The variables that couple subsystems together are not always the variables of primary interest to the disciplinary community developing the submodel. Furthermore, in practice, models as simplified representations of reality usually reproduce phenomena of central interest to their developers better than they reproduce ancillary phenomena. However, coupling parameters that are viewed as ancillary from a disciplinary perspective are crucial to the quality of coupled simulations. Table 1 shows a subset of vegetation parameters used in Dickinson's model BATS, which simulates exchanges of radiation, water, heat and momentum between the atmosphere and the land surface (Dickinson, 1992). Models like BATS do not simulate the parameters in Table 1, rather they are specified. Thus, land cover characteristics cannot change as climate changes or in response to normal, abnormal or extreme weather. Ecosystem models (Parton et al., 1987; Running, 1991) can simulate the dynamics

Table 1.   Selected vegetation/land cover parameters for the BATS model (adapted from Dickinson, 1992).

Parameter
Maximum fractional vegetation cover
Soil depth (total)
Soil depth (surface horizon)
Rooting depth
Vegetation albedo (above and below 7 mm)
Minimum stomatal resistance
Maximum Leaf Area Index
Minimum Leaf Area Index
Stem and dead matter index

of vegetation within and between years. These models, however, were designed to simulate primary production and net ecosystem production on annual time steps. In general, their representation of leaf area index or seasonal biomass is quite poor (Parton et al., in press) compared to the annual values for carbon gain or storage. Thus, the key variables for representing seasonal variation of the biosphere—key to land surface–climate interactions and that BATS and related models are very sensitive to—are simulated very poorly. Significant errors would propagate if these models were coupled, even though both models separately could be validated with great precision, given precise specification of boundary conditions. Some important coupling parameters are not even simulated in many ecosystem models, such as canopy height and fractional canopy cover.

Conclusions

The activity of developing integrated models often leads to the identification of substantial new research, as mismatches between model outputs and model requirements are identified. The traditional way of resolving such problems—increasing detail—is often not feasible due to either computing requirements or the unavailability of suitable data. In such cases, the scientific challenge of seeking new and useful generalizations must proceed, but on a question motivated by the integrated modeling challenge, rather than by a disciplinary concern. Even the identification of the coupling parameters between human and biophysical models will be a challenge, although the considerable effort expended on simulations of productivity of forests, grasslands and croplands, as affected by management, climate and pollutants, provides a foundation in an uncoupled sense.

Several important points emerge in conclusion:

1. The ability to model the effects of human and ecological processes on the physical environment and the recognition of the significance of such impacts on physical climate are new factors that must become an increas-

ingly important aspect of coupled modeling. Climate cannot be treated as an exogenous variable at any scale.

2. The interpolation of information to the finest resolution required in an analysis remains a serious problem. I illustrate this for climate, but similar issues arise with respect to all processes included in IRMs. Nonlinear interactions make this an important concern, since simple averaging of independent variables of nonlinear relationships results in large errors.

3. Disciplinary studies have not always resulted in adequate understanding of the processes that couple subsystems. The poor simulation of coupling variables can result in failures when models that are separately robust for disciplinary applications are linked. When the focus of a study is interdisciplinary, special attention must be paid to understanding and simulation of coupling variables since no coupled model will be better than the simulation of key interfacial variables.

4. Although IRMs are useful in focused studies, regional systems are forced by inputs at their boundaries (matter, energy, decisions) and influence the larger context through their output. While the definition of systems as having strong internal interactions relative to external forcing is a useful fiction for focused studies, it is rarely a valid simplification in the long term. The development of IRMs must proceed in parallel with improvements to models or scenarios of change in the larger context.

Acknowledgments

I was supported by the Climate System Modeling Program of UCAR, funded by the National Science Foundation and the U.S. Department of Energy, and by the National Center for Atmospheric Research, which is sponsored by the NSF, during the writing of this paper. I would like to acknowledge sustained discussion over many years with Tim Kittel on the subject of this chapter and with Fillipo Georgi, Roger Pielke, Dennis Ojima and Bill Parton. Charles Hall provided a thought-provoking and helpful review of this chapter.

References

Aber, J. D., K. Nadlehoffer, P. A. Steudler, and J. M. Melillo. 1989. Nitrogen saturation in northern forest ecosystems—Hypotheses and implications. *Bioscience* **39**:378–386.

Archer, S., C. Scifres, C. R. Bassham, and R. Maggio. 1988. Autogenic succession in a subtropical savanna: conversion of grassland to thorn woodland. *Ecological Monographs* **58**:111–127.

Bretherton, F. B., R. E. Dickinson, I. Fung, B. Moore III, M. Prather, S. Running, and

H. Tiessen. 1992. Report: Linkages between terrestrial ecosystems and the atmosphere. In D.S. Ojima (ed.). *Modeling the Earth System.* UCAR/Office for Interdisciplinary Earth Studies, Boulder Colorado, pp. 181–196.

Dickinson, R. E. 1992. Land surface. In K. E. Trenberth (ed.). *Earth System Modeling.* Cambridge, New York, pp. 149–172.

Georgi, F. 1990. Simulation of regional climate using a limited area model nested in a general circulation model. *Journal of Climate* 3:942–963.

Hall, C. A. S. 1992. Economic development or developing economics: what are our priorities? In M. K. Wali (ed.). *Ecosystem Rehabilitation*, Vol. 1. Policy Issues. SPB Academic Publishing, The Hague, pp. 101–126.

Houghton, J. T., G. J. Jenkins, and J. J. Ephraums (eds.). 1990. *Climate Change: The IPCC Scientific Assessment.* Cambridge University Press, Cambridge, UK.

Langford, A. O., F. C. Fehsenfeld, J. Zachariassen, and D. S. Schimel. 1992. Gaseous ammonia fluxes and background concentrations in terrestrial ecosystems of the United States. *Global Biogeochemical Cycles* 6:459–483.

Parton, W. J., D. S. Schimel, C. V. Coleand, and D. S. Ojima. 1987. Analysis of factors controlling soil organic matter levels in Great Plains grasslands. *Soil Science Society of America Journal* 51:1173–1179.

Parton, W. J., J. M. O. Scurlock, D. S. Ojima, T. G. Gilmanov, R. J. Scholes, D. S. Schimel, T. Kirchner, J-C Menaut, T. Seastedt, E. Garcia Moya, A. Kamnalrut, and J. I. Kinyamario. Observations and modeling of biomass and soil organic matter for the grassland biome worldwide. *Global Biogeochemical Cycles*, in press.

Running, S. W., R. Nemani, and R. D. Hungerford. 1987. Extrapolation of synoptic meteorological data in mountainous terrain and its use in simulating forest evapotranspiration and photosynthesis. *Canadian Journal of Forest Research* 17:472–483.

Running, S. W. 1991. Computer simulation of regional evapotanspiration by integrating landscape biophysical attributes with satellite data. *In Land Surface Evaporation, Measurement and Parameterization.* Springer-Verlag, New York, pp. 359–369.

Schimel, D. S., T. G. F. Kittel, and W. J. Parton. 1991. Terrestrial biogeochemical cycles: global interactions with the atmosphere and hydrology. *Tellus* **43AB**:118–203.

Trenberth, K. E., G. W. Branstator, and P. A. Arkin. 1992. Origins of the 1988 North American drought. *Science* **242**:1640–1645.

Vitousek, P. M. 1990. Biological invasions and ecosystem processes: towards and integration of population biology and ecosystem studies. *Oikos* **57**:7–13.

II

Disciplinary Reviews of Regional Models:
Outsider Perspectives

2

An Overview of Biological Models: A Physical Scientist's Perspective

Bruce P. Hayden

Introduction

In this chapter, I serve as a commentator outside the field or as a gadfly for models developed by ecosystem scientists. My area of interest is the atmospheric sciences. In my view, there are important historical parallels in modeling in the two disciplines. My comparison of these penultimate ecosystem and weather and climate models focuses on common attributes: system heterogeneity, scale, simple structure, memory, succession, state change, and chaos. In order to prevent my Don Quixote tendencies from running rampant, I have reread A. M. Turing's sobering warning about his own model published in his 1952 *Philosophical Transactions* paper:

> This model will be a simplification and an idealization, and consequently a falsification. It is to be hoped that the features retained for discussion are those of greatest importance in the present state of knowledge.

Models are dependent on a state of knowledge and to a significant degree they are ephemeral mathematical constructs. They are always penultimate with the next generation of improved models anticipated. The broad issue here is the merging of ecosystem, physical and social system models at the regional scale. A major research initiative will be required to develop the knowledge and technology to network these diverse kinds of models.

History

Near the turn of the present century, the German hydrodynamist Max Margules used the harsh phrase "immoral and damaging to the character of a meteorologist" to characterize regional weather forecasting (Kutzbach, 1979). To Margules, the regional weather models under development were premature. They were but hypotheses awaiting the bright light of observational data. His work on atmospheric energetics formed the basis upon which the Norwegian School of atmospheric dynamics evolved and became world famous for regional models with weather forecasting utility. By 1933, climate was a dynamical science at the regional scale and two schools of thought vied for intellectual control of dynamic climatology. The Norwegian, Hesselberg, led one school. Bergeron, also a Norwegian, led the other. Atmospheric science was practically a Norwegian science in the 1920s and 1930s.

> Dynamic climatology must be concerned with the quantitative application of the laws of hydrodynamics and thermodyamics . . . to investigate the general circulation and state of the atmosphere, as well as the average state and motion for shorter time intervals.
>
> *Hesselberg (1932)*

> . . . a dynamic climatology should describe the frequencies and intensities of well-defined systems that are more or less closed in a thermodynamic sense.
>
> *Bergeron (1930)*

To Bergeron the atmosphere behaved as a community of thermodynamic entities: air masses, cyclones, and anticyclones with boundaries (e.g., cold and warm fronts) between them. Hesselberg's approach was different. He ignored the entities we recognize as part of day-to-day weather. There were no entities or sharp boundaries in the system he modeled. His approach is the equivalent of ecosystem scientists who see no ecotones—merely a continuous transition from one ecosystem to the next. It has similarities to the spat between the molecular and organismal viewpoints in biology. In ecological parlance, Bergeron was the community ecologist of the climate, and Hesselberg was the ecosystem ecologist. These two approaches to the atmospheric system were simultaneously in vogue and advanced the field of atmospheric science rapidly. Bergeron's "community" approach made its major contribution to weather forecasting and our understanding of climate dynamics prior to World War II. Hesselberg's methods led to the modern, numerical-weather-forecast models of the 1960s, the development of General Circulation Models (GCMs) of climate in the 1970s, and the now famous doubled carbon dioxide runs of these GCM experiments in the 1980s.

The notion that climate is specified by the sum of the frequencies of air mass in a region in a given period is an example of a Bergeron view of a climate. Essenwanger (1954) offers an example of the Bergeron approach. He takes the vapor pressure of the air as the marker of air mass type (Fig. 1A) and illustrates that this aspect of climate is a collective of normally distributed air mass types. The resident of Essenwanger's location would experience a sequence of these thermodynamic entities within a specified period of time. An equivalent ecosys-

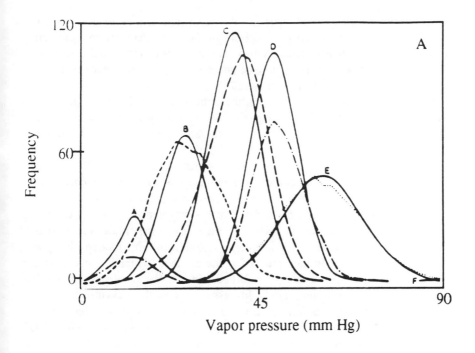

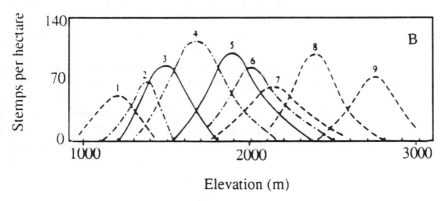

Figure 1. Comparison of climatological and ecological views of the state of the system as the sums of constituent parts. In (A) five types of air masses are specified in terms of their vapor pressures: A = Polar Continental Air; B = Continental and Polar Air; C = Pacific Ocean Air; D = Temperate Maritime Air; and, E = Maritime Tropical Air (after Essenwanger, 1954). Dashed lines are data and solid lines are normal equations for each air mass type. In (B) Whitaker's (1969) tree distribution along an elevation gradient: *1, Vauquelinia californica; 2, Quercus oblongifolia; 3, Q. emoryi; 4, Q. arizonica; 5, Q. hypoleucoides; 6, Q. rugosa; 7, Q. gambelli; 8, Acer grandidentatum;* and *9, A. glabrum.*

tem's view (Whitaker, 1969) specifies the ecosystem at a point in elevation as an assemblage of species (Fig. 1B). Both of these perspectives are community views of the respective systems studied. The kinetics of individuals, air masses or species is the focus of these models, and no conservation of mass and energy as represented by individuals is required.

A Hesselberg comparison between the atmospheric and ecosystem sciences would offer H. T. Odum's 1956 paper on the energy cycle of a community and Oort's 1970 paper on the energy cycle of the atmosphere (Fig. 2). Each model starts with input energy from the sun and ends with stored energy and output heat. Each model follows energy and maintains structure (entropy). These models characterize the equilibrium condition, and the approach in each focuses on the kinetics of mass and energy and follows conservation laws.

The dichotomy of approach offered above is between the study of the fluxes of mass and energy that fuel the system or the study of the entities or individuals that make up the system. This dichotomy is found in most modern disciplines (Huston et al., 1988). In the optics of quantum electrodynamics, every photon interaction with every electron is recorded and integrated to mass effects like diffraction of light (Feynman, 1985). In classical optics, the mass effect is thought of as a singular, measurable response of a continuous and bulk radiant energy. We also see this split in ecosystem modeling. Running's (1988) FOREST-BGC or "big leaf" model has no trees, and the Parton et al. (1987) CENTURY model has no grass clumps, but they follow (integrate) energy and biomass through the forests and grasslands without individuals. In contrast, in FORET class models, like quantum electrodynamics, every tree, each year, is an object of attention and when integrated, defines the character of the stand, and therefore, that of the forest.

Given the existence of these two philosophical starting points in the modeling adventure, several other model groundings must be considered: model level, time increment and dimensionality. The union of these three groundings as a classification of model type is shown in Figure 3. The three levels of model defined in that figure are (1) conceptual and definitional models, (2) field and experimental process models, and (3) stochastic and dynamic simulation models. In his famous book on problem solving, *How to Solve It*, Polya (1945) notes the essential role of conceptual and definitional models as a precursor to more "sophisticated" efforts when he implores the modeler to, "Draw a hypothetical figure which supposes the conditions of the problem satisfied in all its parts." It is fashionable today to refer to these depictions as "cartoons," spaghetti or wiring diagrams, and schematics. They are the first order of models. They are the models used to craft higher-order models. The literature on even the most sophisticated numerical models usually has a cartoon or spaghetti diagram as the road map to the details of the formal model. The wiring diagram for the CENTURY model (Schimel et al., 1990; 1991) is typical (Fig. 4).

The second level of models, field and experimental process models, is usually

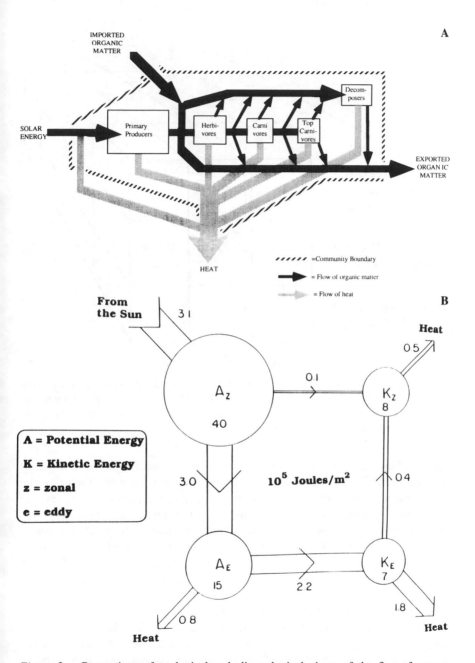

Figure 2. Comparison of ecological and climatological views of the flux of energy through communities and the atmosphere. In (A), a schematic of Odum's (1956) energy flow through a community is shown with solar and organic inputs and dissipative heat as an output. In (B), a schematic of Ort's (1964) characterization of the atmosphere as a dissipative system is shown.

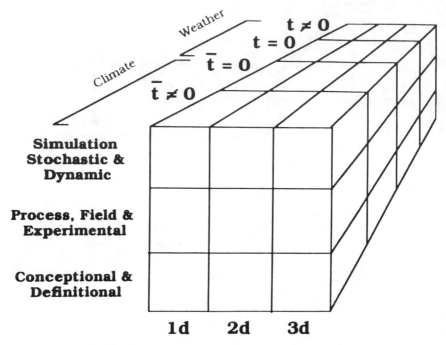

Figure 3. A classification cube of model types. Three types of models are given, each of which may be one, two or three dimensional. In addition, models may be instantaneous ($t=0$), time step incremented models ($t\neq0$), and time averaged (climatologic) or not time averaged (weather).

presented as a mathematical statement, statistical transfer function or curve fit to experimental data. These models are often variable transformation models. Given the magnitude of an input variable, the model calculates the magnitude of the needed output variable. The paper by Greenwood and Walker (1990) offers mathematical statements as models as an approach to problem solving. Burke et al. (1989; 1991) developed a statistical transformation model to go from climate input data to soil carbon storage estimates (Fig. 5). FORET class models (Pastor and Post, 1986; 1988) utilize empirical fits to experimental data to drive or condition higher-order models.

The third level of models, stochastic and dynamic simulation models, generate chronosequences of model output data that track the time history of system properties and thus the dynamics. In the atmospheric sciences, numerical-weather-forecast models and GCMs of climate are the current state of the art. In addition to being explicit in time, these models are spatially explicit in three dimensions. The current generation of ecosystem models are limited in the degree to which they are spatially and temporally explicit. Running (1988) uses the MTCLIM model to generate a high-resolution, spatially explicit climate as a

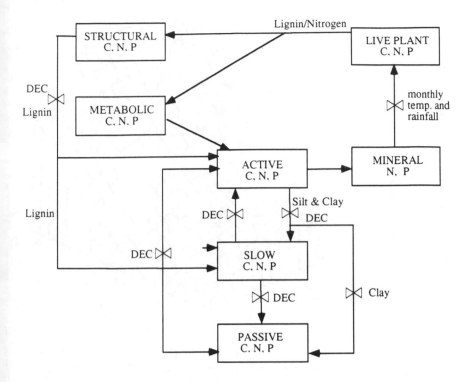

DEC = Climatic Decomposition Parameter

Figure 4. A diagrammatic representation of the CENTURY model (after Schimel et al., 1990; 1991). Soil carbon pools are represented as active, slow and passive and are recharged by decomposition of organic matter.

driver for his FOREST-BGC, "big leaf," model. His model produces two-dimensional, spatially explicit estimates of standing crop and production and decomposition rates.

FORET and CENTURY type simulation models have been point models up to this point, but ZELIG versions are designed to permit interactions between points (stands) and thus become spatially explicit in a limited degree. Other researchers using FORET class models are using seed dispersal as a linkage between points or stands to achieve a measure of spatial explicitness. The development of three-dimensional ecosystem models is some time off.

System Heterogeneity

The focus of most atmospheric and ecosystem modeling is the surface of the Earth. From a regional modeling perspective, this focus is both a blessing and

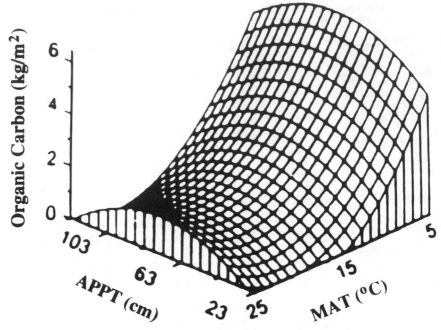

Figure 5. A nomogram for soil organic carbon in the prairies as a function of annual precipitation and mean annual temperature (after Burke et al., 1989; 1991). Nomograms such as Burke's are often used as functional look-up tables in model construction.

a horror. It is a blessing because it is the most interesting and dynamic part of our world. Through the surface layer one finds the greatest fluxes of mass, biomass and energy (Fig. 6). It is a horror because spatial and temporal heterogeneity of most measures of the state of the system are greatest there. Because of this heterogeneity, the surface environment is the most difficult part of the Earth system to model. Heterogeneity becomes smaller as one ascends through the canopy of the biosphere and through the atmosphere. Heterogeneity also becomes smaller with depth downward through soil horizons (Fig. 6). Even the most sophisticated numerical-weather-forecasting models have trouble predicting surface weather conditions. These models do an excellent job of predicting conditions in the free atmosphere, but regression equations are used to translate these forecasts to the surface. These regression solutions out-perform deterministic models that try to capture the physics that gives rise to surface heterogeneity.

Many existing ecosystem models based on the kinetics of individuals are point models and thus ignore spatial heterogeneity. Recently, there have been attempts to merge these point models into a network of points and thus achieve some measure of spatial explicitness (Urban et al., 1987). The critical issues in the application of these models deal with processes that link adjacent points. In

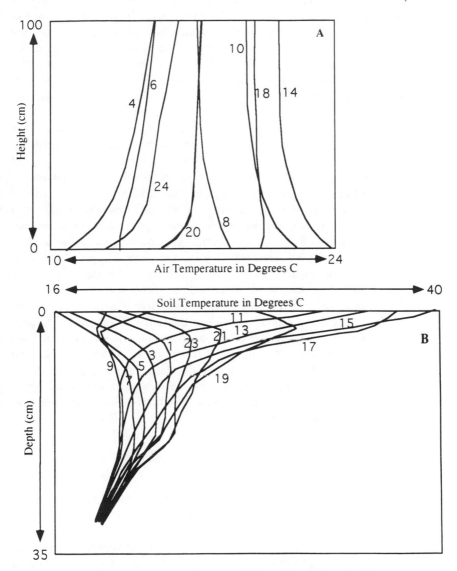

Figure 6. Thermal heterogeneity above and below the surface. In (A), temperature profiles above the surface are given at 4 hr increments during the day. In (B), soil temperature depth profiles are given for 2 hr increments during the day. Thermal heterogeneity is maximum at the surface and decreases with both height and depth.

Urban's work, canopy architecture and its effects on the light geometry at adjacent model points is the linkage that provides spatial explicitness. Other researchers are working on seed dispersal as a mechanism of linkage between independently run point models. These spatially explicit models that use ecological processes such as seed dispersal at adjacent grid points in the model help us learn about ecosystem dynamics, but it is unlikely that they will grow into regional scale, spatially explicit models with predictive capabilities. Topography, groundwater flow paths, climate, and soils and geology will need to be used as boundary conditions that vary from point to point. From the perspective of the focus of this chapter, the connections between ecological, physical and social system model components will pose additional constraints on the approach of networking point models to resolve these problems. Supercomputer capacity and parallel processing will likely be required because a hierarchical system of nested models on both the spatial and temporal bases will be required.

Running (1990; 1991), Burke et al. (1989; 1991), and Costanza et al. (1990) have built two-dimensional, spatially explicit landscape models at the regional scale. Both Running and Burke achieve spatial explicitness in ecosystem production and regional landscape structure by using the observed spatially explicit heterogeneity in landscape and climate to drive their regional models. Costanza's model relies on mass transport of water and sediments between model cells within the Atchafalaya basin of the Mississippi Delta. Both Running's and Costanza's models include surface heterogeneity in the initialization phase. At time zero there is a start-up, spatially explicit landscape that subsequently evolves with each model time-step. Running initializes his model with topography and a leaf area index (LAI) derived from satellite data. Costanza initializes his models with historical aerial photographs and USGS topographic maps. Again the contrast is between models that deal with the kinetics of individuals or with the kinetics of mass and energy fluxes.

Scale

The ideal model includes the essential processes that control the dynamics of the system under study. These processes operate at characteristic spatial and temporal scales and so define the ideal model scales. However, the data available to initialize the model are at the scale of the observational network. The model builder includes those processes that were known, i.e., part of the state of knowledge on the subject, and parametrizes the remaining processes that operate at another scale. The model-programmer is limited by computational capacity and resources, and so these limitations, too, put constraints on the spatial and temporal scales of the model as well as on the geographic extent of the model coverage. The choice of model scale is thus a compromise with costs. O'Neil and Rust (1979) note that simulation error in ecosystem models is related to model scale. In addition, atmospheric GCMs, which use 5° latitude and longitude grid cells, have serious problems at regional and smaller scales. Procedures are used to downscale model output to more useful scales (Dickinson, et al., 1989; and Giorgi and Mearns, 1991). The spatial and temporal resolution and the dynamic variables chosen are not the ones the user usually needs. The user then must either upscale the fine-mesh model output as needed or downscale coarse model output. The problem is especially acute for regional models, since most modeling is either at a larger or smaller scale. Integrated regional models will encounter serious scale problems as the human dimension is included in natural system models.

Modeling approach also plays a role in scale decisions. Where the individual entity is the object of the modeling (Pastor and Post, 1986; 1988), upscaling to meaningful elements of the regional landscape may be the post-mortem requirement of users. Where the models produce aggregate forests or grasslands from the fluxes of mass and energy and the sequestering of carbon (Burke et al., 1991; Running, 1988), users may want downscaling to meaningful landscape elements. GCMs are the penultimate in climate upscaling from meteorological air-parcel theory (Bryson, 1992). Giorgi and Mearns (1991) use "climate inversion" techniques to downscale GCM model output to make it useful for hydrologists and ecosystem scientists. These downscaling requirements and methods' development are called for in Focus 4 of the International Geosphere Biosphere Programme's Biospheric Aspects Hydrological Cycle (IGBP) effort known as the "weather generator" project (Williamson, 1992). Modelers usually use the finest scale possible, while the users want output at the scale they need in their work. Where the landscape is not heterogeneous in structure and composition, the choice of a top-down or bottom-up approach, downscaling and upscaling, is of little consequence or concern.

Simple Structure

Simple structures are the essential components of models. Model simple structures are the theoretical or empirical relationships or equations that permit the input of one set of variables and the output of the desired quantity. Greenwood and Walker (1990) use well-known and accepted equations as simple structures to predict soil productivity and soil-pollution concentrations. Running (1990) uses linear regression to generate equations between annual integrated normalized vegetation index (NDVI) and annual net primary productivity, leaf area index (LAI) (see Fig. 7), and a log linear regression between canopy resistance and the ratio of surface temperature and NDVI. These statistical regression equations, his model simple structures, then are linked in his model. In like fashion, Tucker et al. (1986) find a simple structure regression relationship between global atmospheric carbon dioxide concentrations and NDVI. Burke et al. (1989) solve for

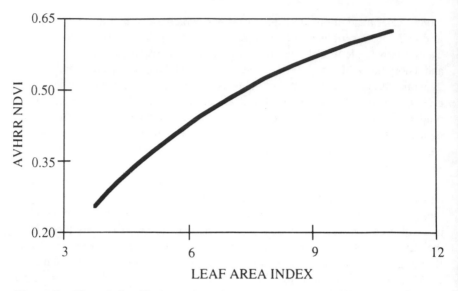

Figure 7. The relationship between leaf area index and AVHRR NDVI is used in Running's 1990 "big leaf" models. The log linear curve shown is a fit to the observational data.

the organic carbon and carbon losses using a traditional trend surface for the variation of soil carbon over annual precipitation and mean annual temperature ranges (see Figure 5). These surfaces are then used to drive soil organic carbon regional model (Burke et al., 1991). Rizzo and Wiken (1992) use a classification of ecoregions based on discriminate function relationships between ecoregion and climate and then used this simple structure model to predict ecoclimate regions based on GCM climate scenarios. In the FORET-type model, realistic but hypothetical functions (Fig. 8) are built to connect the limitation to tree growth to variables like sunlight, growing season degree-days, available nitrogen, etc. (Pastor and Post, 1986). While these parametrizations are crude and based on limited laboratory and field studies, they work well when integrated from year to year. The forms of simple structure used to build ecosystem models do not differ in a fundamental way from those used in the other natural sciences.

Memory

It has long been known that there are direct thermodynamic and dynamic connections between the oceans and the atmosphere. The weather and climate across North America is not independent of the state of the North Pacific Ocean. The state of the ocean depends in part on the weather it experiences. In terms of dynamics, the ocean is somewhat ponderous compared to the atmosphere. The

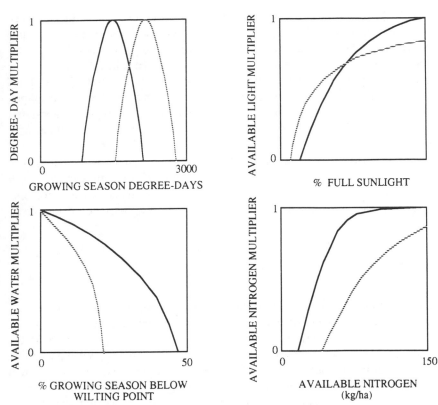

Figure 8. Multiplier functions used in FORET-type models are estimated or measured in lab or field studies. In the four panels (after Pastor and Post, 1986), multiplier functions are shown for Jack Pine (solid line) and Basswood (dashed line).

ocean in this context is a memory element for the atmosphere (Namias, 1972). Ecosystems, likewise, may have parts of their dynamics that function at different speeds. Studies of abandoned fields and in tropical forests clearly indicate that historic farming practices carry forward for decades (Facelli and Pickett, 1990; Hubbell and Foster, 1986). In this case, the soil column retains the impact of past dynamics. These differential speeds of the various dynamical components of a system pose special modeling challenges. For example, atmospheric GCMs have run with much longer time-steps than is required to include most convection and precipitation processes. Traditionally, the practical solution is to parametrize the causal physics of the faster process so that appropriate results at the model speed can be obtained. Parton et al. (1987) and Schimel et al. (1990) developed the CENTURY model (Fig. 4) specifically addressing the fact that parts of the soil–organic-matter dynamical system function on different time scales. One of the central problems of models that include dynamics at more than one time scale is that mass and energy may not be conserved over time. In integrated regional models in which physical ecological and social systems are studied jointly, it will be necessary to incorporate daily, seasonal or annual and extra-annual dynamics into these models. In this case, the political and economic processes may run more slowly than ecosystem processes, and the political and economic system state variables constitute a memory in the system. For example, land clearing and plowing may have ecosystem consequences for decades after the system is let fallow. Clearly, solving the general system memory problem and, more specifically, the problem of "ecological memory," is a research initiative of merit in its own right and may need to be solved before real progress in integrated regional models can be realized. The question of how to incorporate the proper dynamical time scales of the political and economic processes into regional integrated models is critical. These problems are still to be solved within disciplines.

Succession

The notion that current system dynamics depend on past system dynamics, i.e., system memory, is well established in most of the natural sciences. For example, beach morphology determines how the energy of a storm is dissipated on the beach, and the storm, in turn, results in new beach morphology. When the system is near equilibrium and the external boundary conditions that constrain system dynamics are constant, a cyclic system behavior may result. However, where the system is not near equilibrium or where external boundary conditions change, an evolution of the system or a succession may result. Such system progressions may be directional rather than cyclic. The past system-state determines the direction of change. When there is sufficient time, an equilibrium or steady state is approached. Ecosystem models of the FORET type (Pastor and Post, 1986; and 1988) are designed to track such system dynamics in response to disturbance or to changing external conditions like climate (Fig. 9).

Much of the work on models of these types may be classed as perturbation or sensitivity analyses. It is not the dynamics of quasi-steady-state dynamics but rather the approach to some near equilibrium. Here ecosystem sciences are advanced relative to the atmospheric sciences. Ecologists see the manifestation of succession and have observational models against which to judge their models. The atmospheric scientists have a lack of observational models against which to test their "succession" models. Many "shock" doubling of carbon dioxide experiments have been developed and are much touted, but there are no observational models against which to judge them right or wrong. The "transient" or 1% per year increase in carbon dioxide experiments are of great interest because our month-by-month and year-by-year climate history since the middle of the 1800s can be used as the observational model to validate the experimental com-

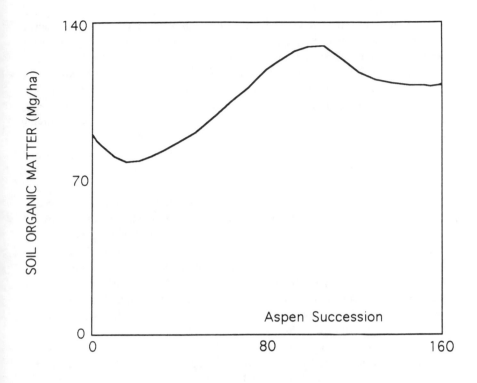

Figure 9. Soil organic matter estimates derived from a FORET-type forest succession model (Pastor and Post, 1986; 1988). Model outputs such as these constitute hypotheses that may subsequently be validated by field studies.

puter model. To date, the models have projected three to five times larger temperature changes than has been realized in nature's historical experiment in carbon dioxide augmentation since the 1850s.

The essential questions are: Can the model predict (1) the direction and pace of successional change, (2) abrupt changes in directions of succession (state change), and (3) the stability of the dynamics of the system when it is at a condition of quasi-equilibrium?

State Change

Of the papers reviewed (see the Introduction) only Costanza et al. (1990) deals with modeling ecosystem state change. Historical aerial photography clearly indicates that the marshland ecosystem exhibited changes from open water to marsh and from marsh to open water over time. While nonlinear models might well exhibit such state change behavior during passage from one chaotic attractor to the next, most linear or linearized models do not. How then do we go from linear models that are succession based to models that capture landscapes that change their state? Costanza used the artificial intelligence approach known as "expert systems" to incorporate system discontinuities or state changes in his model. The expert approach utilizes a series of Boolean if/then-type statements to "force" a change in the state or to require a reinitialization between model time-steps. When certain values of designated state variables are exceeded, the model is reinitialized. If "x" is greater than "p," then change grid cell 233 from marsh to open water. These systems of rules are called "expert systems" because they are are based on experience. Consider the following rule: If the pH falls below 4.8 mobilize aluminum and release it into the water column. If Al exceeds 0.3 micromols in a specified volume, turn off reproduction in top carnivore fishes. These types of rules are a kind of simple structure that are the product of basic research and establish yet another link between the modeling enterprise and laboratory and field studies. Rule-based modeling also raises the issue of the application of fuzzy set theory to model reinitialization. In any case, rule-base modeling depends on a succession-based model that generates model state variables, which when a specified threshold is exceeded, cause the model structure to become discontinuous.

Rule-based models in the atmospheric sciences have diminished over time. Early weather forecasting models were rule-based models operated by forecasters who were well versed in these models. Today weather forecasting is largely deterministic and, to a lesser degree, stochastic. The National Weather Service runs three numerical-weather-prediction models, each twice a day. These three models produce different forecasts, which are sometimes fundamentally at odds. A small group of people study each of the models and applies his or her own "expert system" to say which model is likely to be correct. These experts also draw on the forecast maps, the discontinuities (warm and cold fronts), and the cloud patterns that the computer cannot. These final forecast results are then sent out on the wire as the "hand progs." The children of Bergeron still have their hands in a process that numerically is essentially the continuous field product in the Hesselberg tradition.

Regional models of landscapes and ecosystems will require the exploration of such expert-system-modeling techniques because the history of landscape change and its expected future is the driving force behind the development of regional models. Linear continuous function models will prove inadequate. The ecosystem modeler, like the atmospheric scientist, likely will need the expert-system skills of a naturalist. While expert-systems modeling may well permit the prediction of state changes that are not possible in linear models, nonlinear models are an attractive alternative as a degree of subjectivity can be eliminated. However, nonlinear models may exhibit oscillatory and chaotic behavior that may or may not reflect reality.

Chaos

Most landscape models are built from linear simple structures or are linearized in order to achieve numerically efficient computation. It is unlikely that nature is so linear. In addition to the (CENTURY + FOREST = BCG) FORET-type models of Pastor and Post (1986; 1988) and Bonan (1991), the soil organic matter model of Burke et al. (1991) uses nonlinear empirically simple structures to build their models. The inclusions of these empirically based nonlinearities raise the specter of model instabilities that may or may not be faithful representations of natural system dynamics. The interesting theoretical question is: Do these oscillatory and chaotic perturbations occur in the actual landscape? Do they contribute to spatial heterogeneity? At what scales in space and time do they occur or are there scales at which such instabilities are unrecognizable as implied in the work of Shugart and West (1981)? The notion that natural systems may indeed be chaotic achieved credibility as a result of the computer experiments of Lorenz (1975). Lorenz built simple, three-equation nonlinear models of atmospheric convective systems (Lorenz, 1975) that exhibited a quasi-periodic transitive, chaotic behavior called the butterfly attractor. This behavior traces out a figure in phase space in the shape of a butterfly (Fig. 10). Each wing of the butterfly represents a quasi-stable state of the model. More recently, Lorenz

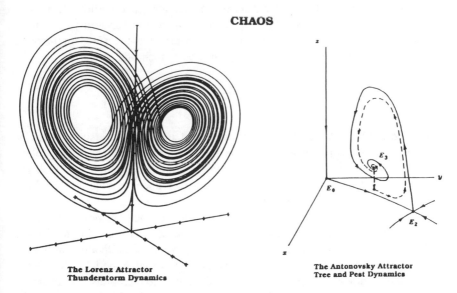

CHAOS

The Lorenz Attractor
Thunderstorm Dynamics

The Antonovsky Attractor
Tree and Pest Dynamics

Figure 10. Attractors in the chaos sense are shown for the Lorenz Attractor and the Antonovsky Attractor. In the Lorenz Attractor the chaotic system is from a parameterization of thunderstorm dynamics, while Antonovsky's attractor is from a set of nonlinear equations representing tree and pest dynamics.

(1984) has generated a simple nonlinear model of the general circulation of the atmosphere and has found similar chaotic behavior. Lorenz concluded that the time limit of weather prediction is about 14 days, and that no weather prediction was possible for longer periods of time. The question of the limits of climate prediction is now undergoing similar analyses.

Lorenz also found that very small changes in initialization of his model resulted in dramatic differences in model results. His systems were heavily dependent on initial conditions. The limits to prediction arise because small errors at one model time-step may become larger at the next time step and so on until the error dwarfs the predictive utility of the model. With the exception of the pioneering work by O'Neil and Rust (1979), the predictive limits of ecosystem models are little studied, but there are indications that problems like those encountered by Lorenz are likely. Antonovsky et al. (1987) constructed a second-order differential-equation forest–pest model that exhibits Lorenzian properties of oscillatory and chaotic behavior (Fig. 10). While we tend to use linear statistical and analytic methods in our analyses of observational and experimental data and we tend to focus on bivariate relationships, it is likely that nature is more nonlinear than we like to admit. If such were the case, complex systems behavior would be expected.

Most of the ecosystem models that use statistical transfer functions or curves fitted to observational or laboratory data fail to include the variances about these functional solutions. Where the variances are oscillatory and abrupt, changes in model state variables are probable. Research on the role of variability or "noise" in model output dynamics is rudimentary, at this point, but it can be demonstrated that chaoslike behavior can be produced in robust stochastic models.

Conclusion

While the fundamental split between kinetics of individuals versus kinetics of mass and energy is common to all systems modeling, integrated regional modeling may, of necessity, have to merge the two types of modeling approaches. Colligative properties from one part of the energy kinetics model may be used to drive another part of the model that tracks and integrates individuals over time. In such merged models, top-down or bottom-up approaches are possible. In the first, the colligative property models provide the boundary conditions for the individual-tracking models. In the bottom-up approach, the colligative properties of integrations of tracked-individuals drive the state variable models. During the last decade, computer computational capacities have grown rapidly and individual-tracking models can now track very large numbers of individuals.

Integrated regional models must be spatially explicit in two if not three dimensions. The spatial relationship between individual matters and regional structure arises in part from interactions between individuals. In addition, the spatial structure of system state variables determines the directional fluxes of mass, energy, information and propagules. Integrated regional modeling will need a field theory of sorts. While field theories are instrumental in weather and climate and in econometric models, they are not well developed in ecosystem models and will be critical in integrated regional modeling.

Integrated regional models will need to be tested for predictive skill at the regional scale. GCMs are generally said to be quite good at the continental and hemispheric scales, but are unreliable within regions. Skill at this scale is not yet demonstrated. Test of skill at the regional scale will require observational data within the region on long time scales. As such, long-term integrated regional studies will be required to develop the needed model simple structure as well as test the models. Weather and climate prediction models usually are tested against simpler models. Three simple predictions usually are offered as standards to beat: long-term means as a forecast (next year will be like the average year), statistical persistence as forecasts (next year will be like the last year but more nearly normal), and detrended statistical persistence as forecasts (next year will be like last year but more like the trended normal). A model must produce realistic results and must do better than chance.

The goal is to specify the structure and dynamics of a system and to make predictions about system states at future times. This will be the goal of integrated regional modeling. To do this, we must develop the simple structure upon which models can be crafted, incorporate the essential spatial and temporal heterogeneity, determine the stability of the model dynamic, and determine the predictive skill relative to a simpler predictive base. These are the research tasks of integrated regional modeling and if negotiated properly much of Turing's (1952) concern posed at the beginning of this paper will be assuaged:

> This model will be a simplification and an idealization, and consequently a falsification. It is to be hoped that the features retained for discussion are those of greatest importance in the present state of knowledge.

References

Allen, T. F. H., and J. F. Koonce. 1973. Multivariate approaches to algal stratagems and tactics in systems analysis of phytoplankton. *Ecology* **54**:1234–1246.

Antonovsky, M. Y., R. A. Fleming, and Y. A. Kuznestov. 1987. The response of the balsam fir forst to a spruce budworm invasion: A simple dynamical model. WP-87-07nl, IIASA.

Bergeron, T. 1930. Richtlinien einer dynamischen Klimatologie. *Met. Zeit.* **47**:246–262.

Bonan, G. B. 1989. A computer model of the solar radiation, soil moisture, and soil thermal regimes in boreal forests. *Ecological Modelling* **45**:275–306.

Bonan, G. B. 1991. Atmosphere-biosphere exchange of carbon dioxide in boreal forests. *Journal of Geophysical Research* **96**(DA):7301–7312.

Bonan, G. B. 1992. Effects of boreal forest vegetation on global climate. *Nature* **359**:716–718.

Bryson, R. A. (1992). A macrophysical model of the Holocene intertropical convergence and jetstream position and rainfall for the Saharan region. *Meteorological and Atmospheric Physics* **47**:247–258.

Burke, I. C., Yonker, W. J. Parton, C. V. Cole, K. Flach, and D. S. Schimel. 1989. Texture, climate and cultivation effects on soil organic matter content in U.S. grassland soils. *Soil Science Society of America Journal* **53**:800–805.

Burke, I. C., T. G. F. Kittel, W. K. Lauenroth, P. Snook, C. M. Yonker, and W. J. Parton. 1991. Regional analysis of the central Great Plains. *Bioscience* **41**:685–692.

Costanza, R., F. H. Sklar, and M. L. White. 1990. Modeling coastal landscape dynamics. *Bioscience* **40**:91–107.

Dickinson, R. E., R. M. Errico, F. Giorgi, and G. T. Bates. 1989. A regional climate model for the western United States. *Climate Change* **15**:383–422.

Essenwanger, O. 1954. Neue Methoden der Zerlegung von Haufigkeits-verteilungen. *Ber. dtsch. Wetterdienst U.S. Zone* (Bad Kissingen) No. 10.

Facelli, J. M., and S. T. A. Pickett. 1990. Markovian chains and the role of history in succession. *Trends in Ecology and Evolution* **5**:27–30.

Feynman, R. P. 1985. *QED: The Strange Theory of Light and Matter*. Princeton University Press, Princeton, NJ.

Giorgi, F. and L. O. Mearns. 1991. Approaches to the simulation of regional climate change—a review. *Reviews of Geophysics* **29**:191–216.

Greenwood, D. J., and A. Walker. 1990. Modeling soil productivity and pollution. *Phil. Trans. R. Soc. Lond. B* **329**:309–320.

Hesselberg, T. 1932. Arbeitsmethoden einer dynamishcen Klimatologie. *Beitr. Phys. Atmos.* **19**:291–305.

Hubbell, S. O., and R. B. Foster. 1986. Biology, chance, history and the structure of tropical rain forest tree communities. In J. Diamond and T. J. Case (eds.) *Community Ecology*. Harper & Row, New York, pp. 314–330.

Huston, M., D. DeAngelis, and W. Post. 1988. New computer models unify ecological theory. *Bioscience* **38**:682–691.

Kutzbach, G. 1979. *The Thermal Theory of Cyclones*. American Meteorological Society, Boston, MA.

Lorenz, E. N. 1975. Nondeterministic theories of climatic change. *Quaternary Research* **6**:495–506.

Lorenz, E. N. 1984. Irregularity: a fundamental property of the atmosphere. *Tellus* **36A**:98–110.

Namias, J. 1972. Experiments in objectively predicting some atmospheric and oceanic variables for the winter of 1971–1972. *Journal of Applied Meteorology* **11**:1164–1174.

Odum, H. T. 1956. Efficiencies, size of organisms and community structure. *Ecology* **37**:592–597.

O'Neil, R. V., and B. Rust. 1979. Aggregation error in ecological models. *Ecological Modeling* **7**:91–105.

Oort, H. H. 1970. The energy cycle of the earth. Scientific American **223**:54–63.

Parton, W. J., D. S. Schimel, C. V. Cole, and D. S. Ojima. 1987. Analysis of factors controlling soil organic matter levels in great plains grasslands USA. *Soil Science Society of America Journal* **51**:1173–1179.

Pastor, J., and W. M. Post. 1986. Influence of climate, soil moisture, and succession on forest carbon and nitrogen cycles. *Biogeochemistry* **2**:3–27.

Pastor, J., and W. M. Post. 1988. Response of northern forests to carbon dioxide induced climate change. *Nature* **334**:55–58.

Polya, G. (1945). *How to Solve It: A New Aspect of Mathematical Method.* Princeton University Press.

Rizzo, B., and E. Wiken. 1992. Assessing the sensitivity of Canada's ecosystems to climate change. *Climate Change* **21**:37–55.

Running, S. W. 1988. A general model of forest ecosystem processes for regional applications. I. Hydrological balance, canopy gas exchange and primary production processes. *Ecological Modelling* **42**:125–154.

Running, S. W. 1992. A bottom-up evolution of terrestrial ecosystem modeling theory, and ideas toward global vegetation modeling. In D. S. Ojima (ed.). *Modeling the Earth System.* UCAR/Office for Interdisciplinary Earth Sciences, Denver, Co. pp. 263–280.

Running, S. W. 1991. Computer simulation of regional evapotranspiration by integrating landscape biophysical attributes with satellite data. In T.J. Schmugge and Jean-Claude Andre (eds.). *Land Surface Evaporation*, Springer Verlag, NY, pp. 359–369.

Running, S. W., R. R. Nemani, D. L. Peterson, L. E. Band, D. F. Potts, L. L. Pierce, and M. A. Spanner. 1989. Mapping regional forest evapotranspiration and photosynthesis by coupling satellite data with ecosystem simulation. *Ecology* **70**:1090–1101.

Schimel, D. S., T. G. F. Kittel, and W. J. Parton. 1991. Terrestrial biogeochemical cycles global interactions with the atmosphere and hydrology. *Tellus Series A* **43A-B**:188–203.

Schimel, D. S., W. J. Parton, T. G. F. Kittel, D. S. Ojima, and C. V. Cole. 1990. Grassland biogeochemistry links to atmospheric processes. *Climate Change* **17**:13–26.

Shugart, H. H., Jr., and D. C. West. 1981. Long term dynamics of forest ecosystems. *American Scientist* **69**:647–652.

Sklar, F. H., R. Costanza, and J. W. Day. 1985. Dynamic spatial simulation modeling of coastal wetland habitat succession. *Ecological Modelling* **29**:261–281.

Tucker, C. J., I. Y. Fung, C. D. Keeling, and R. H. Gammon. 1986. Relationship between atmospheric carbon dioxide variations and a satellite-derived vegetation index. *Nature* **319**:195–199.

Turing, A. M. (1952). The chemical basis of morphogenesis. *Philos. Trans. Royal Soc., London, Ser. B* **237**:37–72.

Urban, D. L., R. V. O'Neill, and H. H. Shugart. 1987. Landscape ecology: A hierarchical perspective can help scientists understand spatial patterns. *BioScience* **37**:119–127.

Whitaker, R. H. 1969. Evolution of diversity in plant communities. In *Diversity and Stability in Ecological Systems*. Brookhaven Symposium on Biology. 22.

Williamson, P. 1992. *Global Change: Reducing Uncertainties*. IGBP, Stockholm, Sweden.

3

An Ecologist's Encounter with Some Models in the Social Sciences

Stephen W. Pacala

Introduction

The development of integrated regional models will obviously require intensive flow of information among modelers in the biological, physical and social sciences. This chapter outlines the short but intensive experience of one ecological modeler with a series of nine modeling papers in the social sciences. The papers were selected by the organizers of the conference to represent a broad range of modeling efforts targeting environmental issues. By design, the collection sacrifices depth for breadth of coverage.

Because I am not a social scientist, the analyses and impressions offered here are undoubtedly less sophisticated than analyses by experts. But this is the point of the exercise. We must communicate accurately across disciplines if we are to achieve the synthesis embodied by integrated regional models. The first step toward that synthesis is to expose the gaps in cross-discipline understanding.

The remainder of this chapter is divided into three sections. In the first, I set the context by outlining a typology of current ecological models. In the second, I discuss the nine social sciences papers in four groups: (A) a grab bag containing Bostrom et al. (1993) and Marknsen (1986), (B) models forecasting future levels of anthropogenic emissions of CO_2 by Edmonds and Reilly (1983) and Reilly et al. (1987), (C) cost-benefit analyses of controls on air pollution by Nordhaus (1991) and Portney and Krupnick (1991), and (D) fledgling integrated regional models by Southworth et al. (1991), Grossman (1991) and Lee et al. (1992). Finally, I draw on the first two sections to offer a series of recommendations about the most productive avenues for future research on integrated regional models.

Setting the Context

One distinction between ecology and some other areas in the biological sciences is that ecological phenomena are generally quantitative. Whereas molecular biologists are typically interested in the structure and qualitative function of genes, ecologists seek to explain numbers, such as the population densities of species or the amounts of organic and inorganic material in different parts of an ecosystem, and how these numbers change through time and across space. As a result, ecological theories are typically grounded in dynamical models (e.g., systems of differential or finite-difference equations) that describe the processes that govern the numbers.

Although beginning much earlier, the modern modeling tradition in ecology flowered during the 1960s and 1970s. Two schools of modeling developed. The first, as exemplified by the work of Robert MacArthur, sought simple analytically tractable models intended to explain rather than to predict and to expose counterintuitive consequences of simple assumptions. The second, as exemplified by the ecosystem models developed under the International Biological Programme, sought complicated computer-simulation models designed to predict the dynamics of specific ecosystems.

Each school of modeling has produced a mixed bag of success and shortcomings. The architects of simple models produced a remarkable array of hypotheticals that remains the dominant focus of research in many areas of ecology. Results include spectacular discoveries of widespread significance, such as the discovery of mathematical chaos by Robert May (May and Oster, 1976). Even so, by the early 1980s, it became apparent that simple models were better at sharpening questions than providing answers (Simberloff and Boecklen, 1981; Strong and Simberloff, 1981). Modelers had produced multiple explanations for important ecological phenomena, but because of the deliberate lack of system-specific detail in the models, there was simply no obvious way to establish a conduit between most natural systems and the correct simple alternative(s) [see the review in May et al (1987)].

In contrast, complex computer simulators were developed for specific ecosystems during this same period that were capable of predicting dynamics in the field [see, for example, Shugart (1984)]. However, because these models were too complicated to be fully analyzed and understood, they typically did not provide useful explanations. Moreover, in their desire to embrace the complexity of nature, modelers usually included unmeasurable "free" parameters. The problem with complex models containing free parameters is that they may be capable of producing an observed phenomenon for the wrong reason. Hence, the widely cited admonition: "Give me four free parameters and I'll make you an elephant. give me a fifth and I'll make its trunk wiggle." Finally, the cumulative effect of the sampling errors that accompany each estimated parameter grows with the complexity of a model. Thus, even if the goal is prediction rather than explanation and even if no parameters are free, then it is still imperative to construct the simplest possible model.

During the past decade, the two schools of ecological modeling have undergone considerable convergence. Jonathan Roughgarden has described this convergence compactly as "the progression from minimal models of ideas and summarizing models of systems to minimal models of systems." We now have analytically tractable models that explain and predict in a number of crucial areas including epidemiology, animal and plant dispersal, intertidal ecology and plant population dynamics [see the review in May et al. (1987)]. Similarly, computer-simulation models with simple underlying structure are now capable of predicting facets of ecosystem dynamics across much of the surface of the globe [for example, see the discussion of the CENTURY model in the chapter by Burke et al. (Chap. 6)]. These developments are important for two reasons. First, they imply a degree of maturity of ecological models that is essential for the development of integrated regional models. Second, they provide the context for the evaluation of models in the social sciences which follows.

Models in the Social Sciences

A Mixed Bag of Two

Two of the nine papers did not fit conveniently into any category and will be discussed briefly at the outset. First, the paper by Bostrom et al. (1991) begins with the observation that public opinion about an environmental hazard must be evaluated in the context of the public's understanding of the hazard. Because risks vary considerably from place to place, Bostrom et al. (1991) argue that people require substantive knowledge of the scientific processes governing environmental hazards and risks, so that they may tailor assessments to their own circumstances. Thus, we must concentrate on what people know about the *processes creating* risk rather than summary statistics about *levels* of risk.

Bostrom et al. (1991) illustrate one method of assessing scientific understanding of risk by interviewing 24 people about radon. Specifically, they obtained a flow chart summarizing the physical and biological processes that determine the level of exposure to radon and the health risk of exposure (Fig. 1). The chart is complicated; it contains approximately 75 nodes, 14 "level-1" concepts and 48 "level-2" concepts. The interviews were designed to determine how much the public knows about the flow chart. Responses were evaluated using a series of creative and novel measures of completeness, accuracy and specificity. The analysis revealed gaps in knowledge with interesting implications. For example, widespread ignorance that radon decays and flushes from a structure gives many the mistaken impression that remediation is futile.

For me, the most dissatisfying aspect of the paper was that it failed to make an explicit case that knowledge of the flow diagram could improve significantly

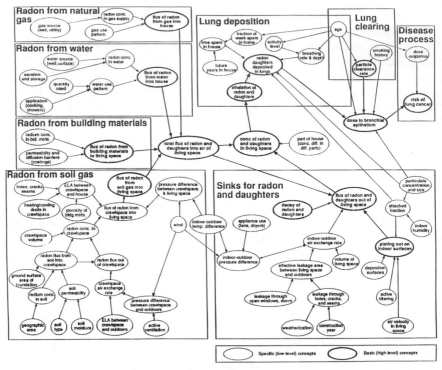

Figure 1. The flow chart describing processes governing residential radon levels and health risk in Bostrom et al. (1992).

one's assessment of risk. The qualitative structure of a dynamical system generally allows only the most limited statements about the system's behavior. In a sense, asking people to assess their level of risk from a knowledge of the flow chart is like asking an ecologist to report the abundances of species from a food web (diagram showing who eats whom).

The paper by Marknsen (1986) investigated factors determining the colonization of metropolitan areas by high-technology companies. Specifically, he used simple linear regression to relate 13 independent variables to the number of high tech plants and jobs in a metropolitan area in 1977, and the changes in these numbers from 1972 to 1977. The independent variables included such factors as housing prices, wages, level of unionization of the work force, airport access, educational options, and local level of university research and development support. He found that the level of business and *amenities* such as climate and educational options were most strongly associated with the level of high tech employment.

This study represents a straightforward application of linear regression to a problem in geography. It shares all of the advantages and disadvantages of any regression study. In particular, as the author acknowledges, one must be careful not to interpret the findings as evidence of cause and effect. For example, the fact that there was a positive association between wages and high tech jobs in 1977 does not imply that either caused the other.

Forecasts of Atmospheric CO_2 Emissions

The list of nine papers included two (Edmonds and Reilly, 1983; Reilly et al., 1987) on models predicting future anthropogenic emissions of CO_2. I will focus on Reilly et al. (1987) because this paper uses a more advanced version of the model (the IEA/ORAU Energy/CO_2 Emissions Model) also studied by Edmonds and Reilly (1983) and others. The two studies cover broadly similar ground, except that the earlier paper also investigates the consequences of policies intended to curtail emissions. Not surprisingly, it comes to the conclusion that unilateral action by the United States would be ineffective, and that effective action would entail either a global reduction in GNP or a global shift toward "CO_2-benign" technologies.

The IEA/ORAU model in Reilly et al. (1987) is a global economic model of extraordinary complexity. It divides the earth into nine geographic regions and models economic growth, population growth, and technological change in each region in relation to the supply, cost and consumption of energy sources from shale oil to nuclear power. The model contains *419 parameters* and forecasts economic conditions, population sizes, and CO_2 emissions through the year 2100.

This complexity is reminiscent of the complexity of ecosystem models developed during the 1960s and 1970s (e.g., summarizing models of systems). Although no parameters appear to be "free," it is difficult to imagine how estimates of such entities as regional labor productivity growth or regional income elasticities through the year 2100 can be other than guesses. In fact, when tested against the historical record, the model actually performs worse than some previous simpler models (Nordhaus and Yohe, 1983). Although the Norhaus and Yohe (1983) model was better able to predict historical variability of CO_2 emissions from 1960 to 1980 than was a simple statistical extrapolation of earlier trends, the Reilly et al. (1987) model was actually less accurate than the statistical extrapolation. Apparently, some degree of mechanistic structure is advantageous, but the complexity associated with too much detail is counterproductive. A passage in Reilly et al. (1987, p. 22) illustrates this message: "They [Nordhaus and Yohe] go on to argue that the fact that their model produces error bounds less than the dynamic historical estimates indicates the improvements in estimates due to careful structural modeling over a naive extrapolation of trends. If we accepted this interpretation, it would indicate that we actually did worse by having an even more detailed structural model."

Why include so much complexity in a model? The obvious answer is the hope that a complex model will somehow allow one to come to grips with a complex problem. A colleague who produces models which forecast levels of toxic waste provides an alternative explanation that is particularly relevant to contentious problems of policy. When I asked him why his models were so complex, he answered: "To protect myself in court." He deliberately builds complexity into models so that when cross-examined by someone hostile to the predictions of the model, he can affirm that the model does indeed account for factor X. If this problem is widespread, then it is imperative that we further educate policy makers about the pitfalls of model complexity.

The most interesting and useful facet of the Reilly et al. (1987) study is its use of Monte Carlo methods to assess the uncertainty in the model's predictions. Briefly, to complete an "uncertainty analysis" of a model, one first obtains as many estimates as possible of the model's parameters from the literature. Second, one constructs a probability distribution for each parameter from the empirical distributions, and draws a large number of parameter sets (i.e., 100) randomly from these estimated distributions. Third, one runs the model for each parameter set to generate an uncertainty distribution of the model's prediction. It is important not to confuse this uncertainty distribution with the sampling distribution used to construct confidence limits in statistics. There is no guarantee that literature values of parameters represent unbiased samples and subjective decisions about the validity of each estimate in the literature accompany the construction of the probability distributions used in an uncertainty analysis (Reilly et al., 1987). In the final step of an uncertainty analysis, one regresses the model's predictions against the parameter values used in each run to determine how much of the variation in the predictions is caused by the "uncertainty" in each parameter. Note that results obtained in this last step will depend both on the sensitivity of the model's predictions to changes in each parameter and on the amount of variation among parameter values reported in the literature.

The distribution of predicted annual CO_2 emissions (Fig. 2) shows enormous uncertainty in the model's predictions. Fully 10% of the runs predicted annual emissions of more than 65 GT or less than 2 GT by the year 2050. Over 50% of this variation was caused by uncertainty in only two factors—the rates of growth of labor productivity and energy efficiency. The implication is that we have very little ability to make long-term predictions of CO_2 emissions, because we simply cannot predict long-term economic or technological change. Because the cumulative effect of estimation errors may grow with the complexity of a model, it is possible that the complexity of the model in Reilly et al. (1987) contributes substantively to the uncertainty evident in Figure 2.

The model also predicted future atmospheric concentrations of CO_2. Over 50% of the runs produced concentrations of less than 600 ppm by the year 2100 (present concentration is approximately 350 ppm). Reilly et al. (1987) stress that this prediction represents a more than 50-year postponement of the doubling time for atmospheric CO_2 predicted by earlier models. While this interpretation is interesting, another interpretation is that the result is an artifact of the uncertainty

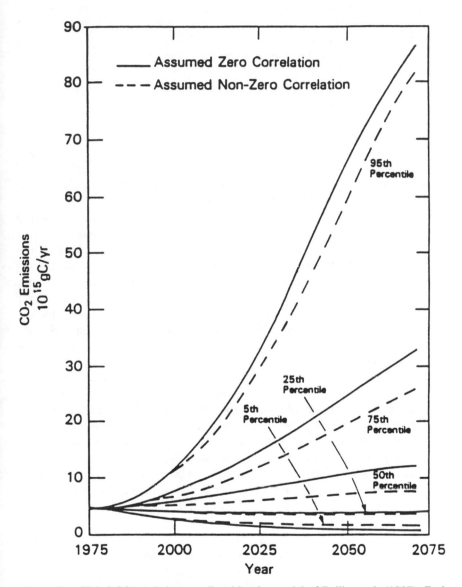

Figure 2. Global CO$_2$ emissions predicted by the model of Reilly et al. (1987). Each curve shows a uncertainty limit. For example, 95% of runs predicted emission levels beneath the curve labeled "95th Percentile." The dashed and solid curves are produced by two marginally different versions of the model.

of the predictions. For example, a model with virtually no predictive power might forecast very low emissions owing to global economic collapse in 50% of runs and very high emissions owing to rapid economic growth in 49% of runs, with the remaining 1% intermediate. Clearly, one should not infer too much from the fact that the median run predicts little change in emissions in this hypothetical example.

Cost-Benefit Analyses of Air Pollution

Two of the nine papers employed the contentious methodology of cost-benefit analysis. Nordhaus (1991) analyzed the costs associated with reducing anthropogenic emissions of CO_2 and other greenhouse gases in relation to the benefits of reduced climate change, while Portney and Krupnick (1991) investigated the costs and benefits of controlling urban air pollution.

Nordhaus (1991) is unique among the nine papers in that it is the only study employing an analytically tractable model. This model is composed of two coupled submodels: (1) a simple, but sensible, pair of equations summarizing changes in atmospheric greenhouse gases and global mean temperature and (2) a simple economic model (model of future per capita consumption) that includes the costs of controlling emissions and the costs of damage caused by climate change. From the model, Nordhaus calculates an elegant expression for the optimal strategy in which the marginal cost of further reduction in emissions equals the marginal benefit of the reduction. He then produces admittedly crude and highly speculative estimates of the elements in this expression.

Estimates of costs are detailed in another study by the author (Nordhaus, 1991b). While by no means certain, these estimates are likely to be far more accurate than are estimates of benefits (benefits are defined in terms of climate-change-induced damages averted by a reduction in emissions). In the author's words: "We now move from the *terra infirma* of climate to the *terra incognita* of the social and economic impacts of climate change" (Nordhaus 1991b, p. 930). Briefly, Nordhaus estimates damages in the United States alone as a fraction of the U.S. GNP, and then uses this fraction to compute global damages. He estimates only damages to marketed goods and services, while acknowledging that "a wide variety of non-marketed goods and services escape the net of the national income accounts [including] human health, biological diversity, amenity values of everyday life and leisure, and environmental quality." Of course, these omitted benefits might (and I suspect do) outweigh damages to marketed goods. Nordhaus acknowledges that his estimates of benefits are, at best, accurate only to within an order of magnitude. Together, the estimates of costs and benefits lead to the conclusion that the optimal steady-state level of emissions is between 70% and 98% of the current levels (see Fig. 3).

The analysis in Portney and Krupnick (1991) is simpler than that in Nordhaus (1991) for two reasons. First, Nordhaus (1991) calculates an optimal *strategy*,

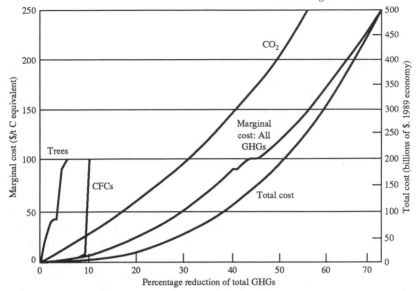

Figure 3. Marginal costs ("curve labeled MC: all GHG's") of reducing anthropogenic carbon emissions and three different estimates of damage caused by climate change from Nordhaus (1991).

whereas Portney and Krupnick (1991) assess the costs and benefits of compliance with only a single course of action (compliance with EPA atmospheric ozone standards). Second, there is a substantial time lag between greenhouse gas emissions and changes in global temperature (which affects the present value of damage caused by current emissions), whereas the response of local ozone concentrations to the level of local emissions effectively is instantaneous.

Like Norhaus, Portney and Krupnick acknowledge enormous uncertainties in their estimates of both costs and benefits, with greater uncertainty on the side of benefits: "Perhaps it is not surprising that uncertainties are greater concerning the benefit estimates presented here" (Portney and Krupnick 1991, p. 252). The benefits are calculated solely from: (1) epidemiological and clinical studies on the increased medical expenditures encumbered because of air pollution levels in excess of the EPA standards and (2) an estimate of the monetary "value" that people place on the increased risk of death associated with air pollution in excess of EPA standards. Thus, the calculation of benefits omits the value of ecological damages as well as many of the amenities afforded by cleaner air. How much would people be willing to pay to avoid damage to ecosystems and the physical and aesthetic discomfort caused by air pollution? Together, the costs and benefits included in the study lead to the conclusion that the cost of compliance with EPA standards exceeds the benefits by approximately an order of magnitude (nationally, 8.8 billion dollars in costs and 250–800 million in benefits).

The application of cost-benefit analysis to environmental problems has been widely and vociferously criticized on a number of grounds (Norton 1988; Randall 1988; Rhoads 1990). Commonly voiced criticisms include the following: (1) The costs of regulation are often easier to measure than the benefits, analyses commonly omit dominant benefits. Note that this is clearly the case in both Nordhaus (1991) and Portney and Krupnick (1991). (2) The method of computing the total value of costs or benefits as a simple sum across the affected population raises concerns about equity. Action that damages a small group of people to a significant extent can be balanced by a small per capita benefit distributed over a much larger group, and vice versa. (3) The standard practice of discounting future gains and losses by the real rate of return on investments leads to the paradoxical result that an ecological disaster in the distant future can be offset (in the analysis) by relatively small rates of financial return on a relatively small amount of capital that would avert the disaster if spent appropriately now. (4) Some [see the discussion in Rhoads (1990)] argue that it is unethical to place a monetary value on human life.

Of these concerns, (1) is most relevant to our discussion here. Because the analyses in Nordhaus (1991) and Portney and Krupnick (1991) omit potentially dominant costs and (especially) benefits, these studies are best interpreted as analogous to the simple exploratory models (e.g., minimal models of ideas) developed by ecologists during the 1960s and 1970s. The studies are useful to sharpen the questions and to reveal the consequences of simple (and perhaps false) assumptions, but they cannot be expected to provide quantitatively accurate conclusions. An ecological model of competition among insect species that omits the dominant effects of predators may help ecologists to develop new ideas, but it would be ridiculous to use the quantitative predictions of such a model to design a program for the control of a specific pest. Similarly, the analyses in Nordhaus (1991) and Portney and Krupnick (1991) may help economists to sharpen their questions, but the quantitative conclusions of these analyses should not be used to formulate policy. Nordhaus (1991, p. 937) acknowledges this point: "Notwithstanding these simplifications, the approach laid out here may help clarify the questions and help identify the scientific, economic and policy issues that must underpin any rational decision."

Fledgling Integrated Regional Models

Three of the nine papers (Grossman 1991; Lee et al., 1992; Southworth et al., 1991) describe models or modeling approaches that straddle ecology and the social sciences. All begin with a digitized map of a location, which is stored in a geographic information system (GIS). Southworth et al.'s (1991) model concerns the province of Rondomia in Amazonian Brazil, Grossman (1991) describes studies of alpine regions in Europe, and Lee et al. (1992) describe, in general terms, the kind of methodology illustrated in the other studies. Here, I focus solely on the study of Rondomia, to illustrate the approach with a specific example.

Southworth et al. (1991) model the land use changes in Rondomia that recently have transformed much of the province from tropical forest to degraded pasture. Their model divides a study area into a large number of small rectangular plots, each the size of a single family's holding. Three linked submodels govern temporal changes in the state of each plot. First, a stochastic settlement-diffusion model determines which forest plots are settled, the length of an owner's tenure on a plot, transfer among owners, and plot abandonment. These state transitions are functions of plot size, agricultural suitability as determined by soil surveys, distances to markets along roads, and the states of other plots in the local neighborhood. Although calibration is currently in its early stages, the settlement-diffusion submodel is designed specifically so that parameters can be measured, and the necessary information is currently being gathered. For an example of a similar model in a more advanced stage of development, see the land use model in Parks (1991). This model is interesting because it is calibrated largely using remote (satellite) imagery and includes land-state transition probabilities that depend on the market prices (i.e., prices of timber, livestock and crops). I suspect that both of these attributes will be added ultimately to the model of Southworth et al., given the considerable efforts to analyze remote imagery of Rondomia (T. Janetos, personal communication) and the obvious importance of agricultural and timber prices to development of the region.

The second submodel in Southworth et al. (1991) governs the land use changes within a plot. For example, the average practice following initial settlement is to clear 7 ha of forest per year and convert this to annual crops. After one or two years in annuals, a field is converted to perennial crops, and then ultimately to pasture.

The final submodel is ecological—it predicts changes in the amount of organic carbon stored on each plot as a function of land use. For example, under average conditions, carbon stocks present in pristine forest decline by over 90% in the first 15 years following clearing. Clearly, one could expand the ecological submodel to predict other characteristics of the ecosystem, including biodiversity. The impetus for an initial focus on carbon probably stems from the inability of global models of the carbon cycle to account for the fate of approximately 2 GT per year of anthropogenically released carbon (Tans et al., 1990). Suspicion that much of this "missing carbon" is being sequestered by ecosystems in the wet tropics [see the recent analysis of a global general circulation model by Sarmiento and Rayner, in prep.], confirms the critical need for the more detailed analyses afforded by regional models such as Southworth et al. (1991). This illustrates how regional models may be integrated into models targeting even larger scales.

Collectively, the three submodels in Southworth et al. (1991) provide spatially explicit predictions of regional population density, land use, land cover and carbon storage, caused by social, economic and ecological processes (see Figure 4). The model also allows one to investigate the consequences of alternative land-use practices. For example, the model predicts that "sustainable" agricultural prac-

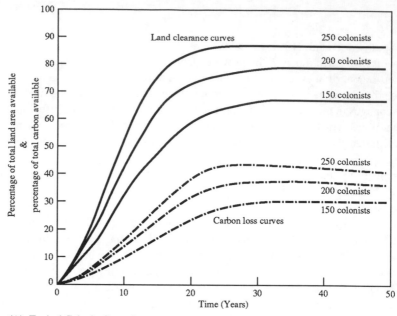

(A) Typical Colonist Scenario

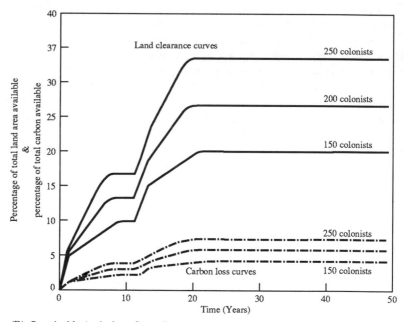

(B) Sustainable Agriculture Scenario

Figure 4.   Levels of land clearance and carbon loss predicted by the model of Southworth et al. (1991) for the province of Rondomia in Amazonian Brazil. (A) Predictions for the current pattern of land use. (B) Predictions if the sustained land use of Japanese immigrants in eastern Brazil were adopted in Rondomia.

tices of Japanese immigrants in eastern Brazil would reduce carbon losses in Rondomia by approximately 80%.

In general, the models described in Southworth et al. (1991), Grossman (1991), and Lee et al. (1992) are closer in spirit to ecological minimal models of systems than they are to minimal models of ideas or summarizing models of systems. The models share a relatively simple underlying structure that facilitates analysis and understanding, and yet are intended to make quantitatively accurate predictions about real economic–social–ecological systems.

Conclusions

The success of any integrated regional model (or any complex model) will depend not only on how well the model predicts, but also on how much the model explains. Typically, explanatory power decreases as complexity increases. Because predictive power may also decrease as complexity increases, it is imperative that we work to develop compact integrated regional models.

An efficient strategy for developing explanatory and predictive integrated regional models is obviously to focus initially on problems at the intersection of relatively mature ecological and social models. Thus, I suspect that successful models will not include economic cost-benefit assessments of environmental and social changes, because a consensus has not yet been reached about how to measure and combine important components of both costs and benefits (Nordhaus, 1991; Portney and Krupnick, 1991). Similarly, I suspect that successful models will not address problems requiring long-term and large-scale forecasts of economic growth. The models of future carbon emissions reviewed here illustrate that such forecasts remain unmanageably complex and intractable (Edmonds and Reilly, 1983; Reilly et al., 1987). In contrast, it appears that integrated regional models of the ecological and social processes controlling land use, vegetation and ecosystem dynamics are ripe for development (Grossman et al., 1991; Southworth et al., 1991). Perhaps these could be coupled with the kind of regional climatic models described on the paper by Schimel (Chap. 1).

This is not to say that modeling problems outside the intersection of mature disciplinary models should be ignored. Rather, such problems should remain temporarily the domain of theoreticians interested in exploring the range of possible interactions among biological, physical and social processes.

References

Social Discipline

Bostrom, A., B. Fischoff, and M. G. Morgan. 1992. Characterizing mental models of hazardous processes: A methodology and application to radon. *Journal of Social Issues* **48**:85–100.

Edmonds, J., and J. Reilly. 1983. Global energy and CO_2 to the year 2050. *The Energy Journal* **4**:21–47.

Lee, R. G., R. Flamm, M. G. Turner, C. Bledsoe, P. Chandler, C. DeFerrari, R. Gottfried, R. J. Naiman, N. Schumaker, and D. Wear. 1992. Integrating sustainable development and environmental vitality: A landscape ecology approach. In R.J. Naiman (ed.). *Watershed Management: Balancing Sustainability and Environmental Change.* Springer-Verlag, New York, pp. 497–518.

Grossman, W. D. 1991. Model- and strategy-driven geographical maps for ecological research and management. In P.G. Risser and J. Mellilo (eds.). *Long Term Ecological Research: An International Perspective, Scope 47.* John Wiley & Sons, New York, pp. 241–256.

Marknsen, A. 1986. Where and why high tech locates. In A. Galsmeier, P. Hall, and A. Marknsen (eds.). *High Tech America.* Allen & Unwin, Boston, pp. 144–169.

Nordhaus, W. D. 1991. To slow or not to slow: The economics of the greenhouse effect. *The Economic Journal* **101**:920–937.

Nordhaus, W. D., and G. Yohe. 1983. Future carbon dioxide emissions from fossil fuels. In *Changing Climate: Report of the Carbon Dioxide Assessment Committee.* National Academy Press, Washington, pp. 87–153.

Portney, P. and A. Krupnick. 1991. Controlling air pollution: A benefit-cost assessment. *Science* **252**:522–528.

Reilly, J., J. Edmonds, R. Gardner, and A. Brenkert. 1987. Monte Carlo analysis of the IEA/ORAU energy/carbon emissions model. *The Energy Journal* **8**:1–29.

Southworth, F., V. Dale, and R. V. O'Neill. 1991. Contrasting patterns of land use in Rondonia, Brazil: Simulating the effects on carbon release. *International Social Sciences Journal* **130**:681–698.

and the following:

May, R. M., and G. F. Oster. 1976. Bifurcations and dynamic complexity in simple ecological models. *Amer. Nat.* **110**:573–599.

May, R. M., J. Roughgarden, and S. A. Levin. 1987. *Perspectives in Theoretical Ecology.* Princeton University Press, Princeton, NJ.

Nordhaus, W. D. 1991b. The cost of slowing climate: a survey. *The Energy Journal* **12**:37–65.

Norton, B. 1988. Commodity, amenity and morality: The limits of quantification in valuing biodiversity. In E. O. Wilson and F. M. Peter (eds.). *Biodiversity.* National Academy Press. Washington, D.C., pp. 200–205.

Parks, P. J. 1991. Models of forested and agricultural landscapes: integrating economics. In M. G. Turner and R. H. Gardner (eds.). *Quantitative Methods in Landscape Ecology.* Springer Verlag, New York, pp. 309–322.

Randall, A. 1988. What mainstream economists have to say about the value of biodiversity. In E. O. Wilson and F. M. Pater (eds.). *Biodiversity.* National Academy Press, Washington, D.C., pp. 217–223.

Rhoads, S. E. 1990. *The Economist's View of the World*. Cambridge University Press, Cambridge, UK.

Sarmiento, J. L., and P. Rayner. 1993. An atmospheric transport contraint on anthropogenic CO_2 source and sink distributions. (Manuscript in preparation).

Shugart, H. H. 1984. *A Theory of Forest Dynamics*. Springer Verlag, New York.

Simberloff, D. S., and W. Boecklin. 1981. Santa Rosalia reconsidered: size ratios and competition. *Evolution* **35**:1206–1228.

Strong, D. R., Jr., and D. S. Simberloff. 1981. Straining at gnats and swallowing rations: character displacement. *Evolution* **35**:810–812.

Tans, P. P., T. J. Conway, and T. Nakazawa. 1990. Latitudinal distribution of the surfaces and sinks of atmospheric carbon dioxide derived from surface observations and an atmospheric transport model. *Journal of Geophysical Research* **94**:5151–5172.

4

Uncertainty in the Construction and Interpretation of Mesoscale Models of Physical and Biological Processes

Richard A. Berk

Introduction

It is impossible to do serious modeling in the social sciences without confronting uncertainty. Social scientists rarely have effective control over their research setting, so that the phenomenon being studied cannot be isolated from a host of confounding influences. For much the same reason, social science data usually come with a healthy dose of measurement problems; at the very least, the measures contain substantial "noise." Finally, popular research designs in the social sciences, resting on probability sampling and/or random assignment, necessarily introduce a chance component into any dataset.

It follows that social scientists must invest significant time and resources to reduce uncertainty and to characterize accurately the uncertainty that remains. It also follows that in this effort statistical tools are essential. Because the discipline of statistics has at least one foot in applied probability theory, the quantification of uncertainty is a cornerstone of the enterprise. As Cox and Hinkley observe (1974, p. 1), "statistical methods of analysis are intended to aid the interpretation of data that are subject to appreciable haphazard variability." Or as one popular aphorism puts it, "statistics is never having to say you're certain."

My task was to review several papers from the physical sciences related to integrated regional models (IRMs). What was perhaps most striking about the papers I reviewed was the almost total absence of any systematic discussion of uncertainty. Were the empirical aspects of each paper undertaken under strict laboratory control and with instruments capable of extremely high precision, the neglect of uncertainty might be understandable. But from the perspective of this

Thanks go to Jan Deleeuw for help in explicating the hierarchical linear model.

outsider, the data represented in the papers look suspiciously like the data routinely collected in the social sciences.

The lack of concern with uncertainty (to be defined shortly) is perhaps most explicitly demonstrated in the fascinating paper by Urban et al. (1987). In that paper, a hierarchical model of landscape ecology is proposed on which I will later build. Watersheds are nested within landscapes, stands are nested with watersheds and gaps are nested within stands. But the data envisioned are apparently deterministic.

Equally intriguing, but equally deterministic, are the six empirical papers I was asked to review. To begin, Dickinson et al. (1989) describe a regional climate model for the western United States. Large-scale weather systems are simulated by using GCM (General Circulation Model) output as boundary conditions for mesoscale simulations. The simulations, in turn, are compared to some observed weather patterns. The authors conclude that the coupled mesoscale models perform better than the GCM alone. One wonders whether the conclusions would have been any different had uncertainty been formally included, especially since summary statistics are computed and treated as point estimates. The deeper issue is what is meant by model testing. Unless uncertainty is incorporated in model testing exercises, it is unclear, at least to me, how one distinguishes between being lucky and being right (or between being unlucky and being wrong). Similar problems in statistics are examined under the rubric of "cross-validation" (Picard and Berk, 1990; Picard and Cook, 1984).

Working on related problems, Pielke and his colleagues (1991) consider the ways in which mesoscale landscape spatial variability should be included in large-scale atmospheric models. They conclude that heat fluxes from different topography and land use patterns are too large to be ignored safely. To this outsider, the findings are both interesting and important, but only point estimates are reported.

Trenberth et al. (1988) explore the links between the 1988 drought in North America and the 1986 to 1987 El Niño in the tropical Pacific. They show how large-scale atmospheric circulation anomalies affecting the Pacific Coast of the United States might well be explained by changes in sea surface temperature, that in turn led to a northward displacement of the intertropical convergence zone southeast of Hawaii. Yet, despite an acknowledgment that many of the data-based numbers used in the computations "are subject to considerable uncertainty" (p. 1642), uncertainty is not formally represented in the findings.

Shukla et al. (1990) employ a coupled model of the global atmosphere and biosphere to explore the effects of deforestation on regional and global climate. As interesting as the science in this paper were the broad implications for public policy. But again, the reader is only given point estimates from calculations that no doubt contain considerable uncertainty.

Jacob and Wofsy (1988) simulate the boundary-layer chemistry over the Amazon forest during the dry season. Some of their findings will be considered briefly later, but suffice it to say, that while their results are rich in substantive implications and provide a number of instructive insights, there are no stochastic elements explicitly introduced.

Finally, the review paper by Girogi and Mearns (1991) considers the variety of ways by which regional climate change is explored. It is telling that in this excellent overview of the state-of-the-art, there is no systematic discussion of uncertainty in the sense considered below.

Therefore, I will concentrate on uncertainty in the construction and interpretation of mesoscale models of physical and biological processes. There will be good news and bad news. The bad news is that the amount of uncertainty may be substantially underestimated in the analyses I was asked to review. The good news is that there is much information currently being ignored that could add substantial precision to the findings reported.

So that I am not misunderstood, I should stress that I truly enjoyed reading each of the papers. I learned a great deal and developed some appreciation for the complexity of the problems addressed and for the considerable talents of the scientists involved. Consistent with scientific conventions, however, I will focus on ways in which the research might be improved.

Some Preliminaries: Thinking about Uncertainty

An obvious place to start is with a conception of uncertainty. Unfortunately, there is no single definition about which widespread consensus exists, and even a summary of the issues will rapidly surface a number of heated controversies (Barnett, 1982). For purposes of this chapter, it will suffice to approach uncertainty through an answer to the following question: Could the results of any given study have been different because of *unsystematic* factors beyond the scientist's control? That is, if the study were repeated under conditions that as closely as possible mirrored the conditions of the original study, would the results be different because of "haphazard" variability?

If the answer is "yes," uncertainty exists, and one must consider if the disparity between the original study and the hypothetical repeated study could be large enough to care about. Could the substantive conclusions meaningfully differ? If the answer is "no," the uncertainty can be safely ignored. But if the answer is "yes," it is essential to formally and explicitly characterize the uncertainty.

Restated, scientific results contain important uncertainty if different conclusions could emerge even under an ideal replication attempt; the initial study's results are significantly affected by unsystematic forces. Then, what the scientist takes to be a stable feature of the data may in fact represent a "fluke." Note that systematic errors are *not* included, since they are replicable (i.e., not haphazard). Systematic errors may be more usefully conceptualized as "bias," leading to results that are consistently off the mark in a particular direction. In the articles

I reviewed, the possibility of systematic errors often was addressed, sometimes in a very sophisticated fashion (e.g., via sensitivity analyses). Systematic errors are not, consequently, addressed in this paper. It is entirely possible that by ignoring the impact of systematic errors, I am overlooking a far more serious problem than uncertainty. However, before that judgment can be made, the role of uncertainty in IRMs would need to be formally assessed.

To help fix these ideas, consider the following illustration. Suppose the ocean temperature at some time and place is measured at 30°C. The scientist who will use that measure then undertakes a mental experiment. If the temperature were measured again with the same instrument at the same time and place, would the result be exactly the same? If the answer is "yes," there is no uncertainty. If the answer is "no," and if the new measure could be meaningfully different from the old measure, there is uncertainty that needs to be formally addressed. Note that if the scientist knows that the measure is in error by *exactly* 2° (i.e., the true temperature is 28°C), the measure is biased, but there is no uncertainty. Finally and more realistically, the scientist may *suspect* that the observed measure is 2° too high. That is, were the measure to be taken again at the same time and place, the scientist's "best guess" is that the observed temperature would again be 2° too high, but probably not exactly 2° too high; there would likely be haphazard variability that could not be precisely anticipated. Then, both bias and uncertainty are implicated. The suspected 2° disparity represents the bias (as before), and haphazard variability around that *true* ("best guess") temperature represents the uncertainty.

There are a number of ways in which this could be more formally represented. For example, one might conceptualize the true temperature (i.e., 28°C) as the mean of a normal distribution and conceptualize the uncertainty as the standard devision of that distribution (e.g., 1°). The bias then would be the difference between the observed temperature and the mean of that particular normal distribution.

Data Collection as a Source of Uncertainty

As the illustration immediately above suggests, uncertainty commonly derives from the ways in which data are generated. Measurement is clearly one source. All instruments have limited precision, even under the best of circumstances, in part because of some lower bound of resolution and in part because of various kinds of interference. And when the instrument is pushed beyond the precision it can deliver, uncertainty will be built into the data collected.

Even with *perfect* measurement, however, uncertainty is inherent in any dataset. In science, one is never interested in a given dataset per se, but in what can be learned from that dataset about some larger universe of phenomena. Thus, any scientific dataset is necessarily a sample. For some nonscientific empirical questions, the given data *are* all one cares about. For example, in litigation involving medical malpractice, the empirical question may be what a physician may have actually done during one particular surgical procedure on one particular patient. Generalization to other procedures and/or other patients is then beside the point.

Suppose, for instance, one has several satellite images from which the vegetation biomass for a particular region can be computed. Clearly, a second fly-by several days later could produce different estimates of biomass because the amount of vegetation biomass will not be constant. If the vegetation biomass is prairie grass, for example, grazing of which the scientist is unaware could introduce substantial haphazard variability. That is, were the study done again, the results could be different. In short, because of haphazard variability in virtually all empirical phenomenon, any dataset can be conceptualized as a realization of all possible realizations that could have occurred. The data on hand are but a sample from a population of possible realizations. Therefore, uncertainty is inherent.

Some datasets are also explicitly samples selected by the scientist from some population of interest. For instance, soil moisture may be measured directly by taking soil samples, ideally selected by probability sampling from a sampling frame in which the land surface is divided into a fine-grained grid. Under probability sampling, each cell in the grid would have a known probability of being chosen. The result is a sample from which proper inferences to the population grid can be made (Cochran, 1977). But in addition, new uncertainty is introduced. Were one to do the study again, probability sampling would almost certainly produce a different sample of grid cells and, consequently, different results.

It may be worth noting that probability sampling is unnecessary if the population is sufficiently homogeneous with respect to the variables in question. For example, if data from tethered balloons will generate essentially the same vertical structure of the atmosphere over the Amazon Basin regardless of when during a 24-hr period the balloon is released and regardless of from where in the Basin the balloon is released, probability sampling is not needed. Virtually any sampling strategy will produce the same substantive story.

My assigned reading suggests that probability sampling is actually quite rare in the modeling community. Rather, researchers more commonly work with "convenience samples." A "convenience sample" is data from some population not selected by probability procedures, but by mechanisms that respond to resource constraints. For example, the reason why one might work with weather data from certain locations is that for all practical purposes, such data are only available from places where there happen to be monitoring stations; in this case, the resource constraint is effectively binding. Were the study done again, exactly the same data points would be chosen. Therefore, the explicit sampling undertaken by the researcher is in this instance *not* a source of uncertainty (although perhaps a source of bias).

To summarize, data collection may lead to three sources of uncertainty: (1)

random sampling error produced by probability sampling, (2) random measurement error due to instrument limitations and interference, and (3) inherent haphazard variability in the phenomenon itself. The implications of all three will be considered shortly.

Data Analysis as a Source of Uncertainty

All data analysis requires decisions based in part on judgment. Problems always surface for which there are no clear solutions, and scientific intuition then will lead the research down some paths rather than others. Whether these kinds of decisions are possible sources of uncertainty or bias depends on whether the decisions are replicable. If the study were hypothetically repeated, and the scientist would make the same decision again, that decision is a potential source of bias. If the scientist would perhaps make a different decision, that decision is a potential source of uncertainty. For example, some of the data may look anomalous, but there may be no way to determine if the anomalies reflect a real error or some unexpected aspect of the phenomenon being studied. Often the scientist will have to make an informed guess. If the same informed guess would be made were the study repeated, the guess would be a potential source of bias. If the same informed guess might not be made were the study repeated, the guess would be a potential source of uncertainty.

The mental experiment implied by repeating the study assumes starting the study again each time from the beginning. This abstraction does not permit sequential learning by the scientist. Learning something from one study that would be used in the next violates the conception of a pure replication. And the conception of a pure replication is required, since in the mental experiment being undertaken, it is the credibility of the *single study in question* that is being evaluated. One is asking what would happen were *this study* repeated.

Mistakes of various kinds can be subjected to the same reasoning. If the mistake would be repeated were the study done again, the result would be a bias. If the mistake would not be repeated (or perhaps a new mistake made), the result would be increased uncertainty.

While uncertainty introduced during data analysis is probably more important than commonly acknowledged, it is difficult to formally represent in a reasonable manner. The processes by which scientists analyze data are not well understood and are surely very complex. Consequently, there is no off-the-shelf statistical technology available.

However, it is possible to introduce one form of uncertainty formally in the data analysis if the *scientist responsible for the data analysis* can formally represent his/her *personal* uncertainty within a Bayesian framework. Even an introduction to Bayesian statistics is beyond the scope of this chapter, but perhaps a brief illustration will convey some of the key issues.

In their article "Photochemistry of biogenic emissions over the Amazon Forest," Jacob and Wofsy (1988, p. 1480) write: "Isoprene emissions from vegetation are strongly dependent on temperature and radiation. We assume an exponential dependence on temperature with a coefficient of 0.2 per degree Kelvin . . . ."

The exponential is a special case of the gamma distribution in which one of the two parameters for the gamma is set to 1 (Johnson and Kotz, 1970, p. 166). By instead allowing that parameter (α) to vary around 1, a number of different shapes are produced. Each can be thought of as a competitor to the exponential. Now, suppose that Jacob and Wolfsy were prepared to entertain a number of "exponential like" distributions. They could obtain these distributions by taking a variety of different values for α. Moreover, a suppose *a priori* they were able to express their beliefs about α in the form of a probability density. They might decide to express their beliefs about α as a normal distribution, for instance, with a mean of 1 and a standard deviation of 0.25. Then, this expression of their beliefs about the likely values for α could be introduced formally into the results through a sensitivity analysis.

For example, they might draw at random a substantial number of values of α from a normal distribution with a mean 1 and a standard deviation of 0.25. The result would be a *distribution* of numerical values for the outputs of the analysis. Thus, their graph (Figure 4) showing isoprene emission flux by time of day, would show not a curved line, but perhaps a curved band representing a distribution of emission values for each time of the day. And more formal probability statements reflecting the consequences of the authors' initial uncertainty about α could also be made. For instance, the peak of emission flux at noon could be expressed not as a point, but as the probability that α fell within a certain band.

The illustration from Jacob and Wofsy is instructive because *two* kinds of uncertainty are addressed simultaneously: uncertainty in the value of a given parameter of the model *and* uncertainty in the functional form of the model. Usually, these are distinct sources of uncertainty that are examined with different distributions (or different sets of distributions). The general point is that promising statistical tools exist, at least in principle, to address both kinds of uncertainty (Draper, 1993; Raftery, 1993). However, these tools depend in part on a subjective conception of uncertainty as degrees of belief, which for this paper, will take us somewhat far afield.

Some Implications of Uncertainty

Before turning to a more formal treatment of the implications of uncertainty, corresponding to the kinds of modeling I was asked to review, it may be useful briefly to introduce the basic issues. Perhaps the most important lesson is that results conceptualized as single numbers, rather than as intervals, are likely to be misleading. In particular, computed numbers need to be reconceptualized not as points, but as distributions.

These distributions, in turn, need to be carried through to later computations and conclusions. This seems to be done commonly with potential biases. For example, simulations often are done under varying assumptions, and a range of results reported. The same philosophy needs to be applied to uncertainty.

There are at least three places where the formal consideration of uncertainty should appear. First, all summary statistics from a given dataset and model should be reported with their sampling distributions or some instructive information from their sampling distribution. Formal "confidence intervals" are one example. Confidence intervals, which build on the distributions of summary statistics, are one popular way to communicate quickly the amount of uncertainty associated with any point estimate. For instance, when reporting the precipitation in a region, to that number could be attached a confidence interval (e.g., plus or minus 1 centimeter).

Second, in a similar fashion, all forecasts or predictions need to be presented as predictive distributions, perhaps summarized with confidence intervals. A forecast is really just a summary statistic for data that are not (yet) available. The same logic applies to models tested by examining how well the model's forecasts correspond to the observations in a new data set.

Third, binary statements about whether a particular statement is true or false should be reported along with probability assessments of how likely it is that the statement is true or false. Such assessments are routinely made as part of testing for "statistical significance."

All statements of uncertainty, however, depend on a formulation of how the uncertainty was introduced. That is, one needs a model of the sources of the uncertainty. Sometimes that model is little more than a formal statement of how the data were sampled from some population. But for the kinds of empirical work reported in the papers I read, a far more complicated model is required. I turn to that now.

A Hierarchical Framework for Studying Uncertainty

If there were one thread that could be found in all of the papers I reviewed, it is the necessity of working on more than one level of spatial and/or temporal aggregation. Perhaps the most explicit example was found in the paper by Urban and his colleagues, where a "hierarchical perspective" was explicitly adopted. For example, gaps are nested within stands, which are nested within watersheds, which are nested within landscapes.

Recent developments in the social sciences and statistics have also brought hierarchical models to the fore. The literature is rich (Bryk and Raudenbush, 1992; Maritz and Lwin, 1989) and growing rapidly. While the details are well beyond the scope of this chapter, I will turn briefly to hierarchical models as a way to address, far more formally, the role of uncertainty.

We can begin very simply with a single equation at a microlevel and a single equation at a macrolevel:

$$y_{ij} = \delta_j + e_{ij}$$

$$\delta_j = \delta + \epsilon_j$$

where
y_{ij} = observation i for setting j
δ_j = causal effect for setting j
e_{ij} = random disturbance i for setting j
δ = causal effect at macrolevel and
ϵ_j = random disturbance for setting j

Consider as a very simple illustration that a scientist is interested in estimating the snowpack depth in each of some set of geographical sites. Snowpack depth is to be measured by reading a number on a calibrated rod anchored in the ground at each site. Were several readings taken at a given site, it is almost certain that the readings would differ. The "eyeball" assessments would vary a bit from reading to reading; the measurement process would have inherent haphazard variability. Aware of uncertainty in the measurement process, the scientist decides to take five readings from each rod and compute the mean.

Each reading is symbolized above by y_{ij}, where i represents the particular reading $(1, \ldots, 5)$, and j represents the particular site among many possible sites. The mean of the five readings for site j is represented by j, and random variation around that mean due to measurement error is represented by e_{ij}. The measurement errors are assumed to be independent of one another, with a mean of zero and some variance V_j. It is common to assume that the variance V_j is the same across sites. Then, the subscript is dropped. The zero mean follows if the initial measurement process is assumed to be unbiased; in principle, the measurement error cancels out if a sufficient number of readings are taken. One interpretation of the first equation is that the mean δ_j is an estimate of the true snowpack depth. Hence, one may speak of δ_j as the causal effect for the observation y_{ij}.

At the macrolevel, the scientist believes that the sites differ in their true snowpack depth. These differences across sites are taken to be the result of haphazard variability ϵ_j around a grand mean for snowpack depth δ computed across all of the sites. Sources of the haphazard variability might be topography and prevailing winds, affecting snow drifts, and exposure to sunlight (which influences melting). The resulting variability is haphazard because the scientist has not measured its causes and included them in the formulation, and because, if the study were done again, the true snowpack depths are likely to be somewhat different. Much like for the microlevel, it is common to assume that ϵ_j has a mean of zero and some variance W; the haphazard variability is assumed to cancel out.

Four lessons follow immediately, even from this very simple formulation. First, there are *two* sources of uncertainty, one at the microlevel and one at the macrolevel. Both need to be taken properly into account.

Second, the disturbances at the microlevel are necessarily correlated. Note that if one substitutes the macro equation into the micro equation, all observations within a given site share a common disturbance ϵ_j. Hence their covariance is nonzero. In the snowpack illustration, the source of this covariance within site might include the time of day when the five measures are taken; the timing of the observations will likely be more homogeneous within sites than across sites. Also, a single technician may take all the readings at a given site, but technicians may differ across sites. Conventional statistical procedures will not be able to take this sort of covariance into account and, therefore, misrepresent the uncertainty.

A third lesson is that there is information about each site's snowpack depth to be found in the grand mean for the sites. Note that each site's true snowpack depth is a function of the grand mean. It follows that one can obtain a more precise estimate of each site's true snowpack depth by using not just the computed mean snowpack depth in a site (δ_j), but a weighted average of δ_j and a proper estimate of δ; here, simply the mean of the site means. The weights, in turn, would be determined by the variability of the readings within a site and the variability in the means across sites. For example, if there is substantial haphazard variability in the five readings at a given site, but little haphazard variability in the means across sites, the importance of the site mean would be reduced compared to the importance of the grand mean when the weighted average of the site mean and the grand mean was computed. The underlying principle is that means (and grand means) based on data with greater amounts of haphazard variability are given less weight in the weighted average.

Finally, in order for the hierarchical model just described to be implemented, the requisite data need to be anticipated in the research design. In particular, a number of measures at the micro level (at each site) are required. If only one measure is obtained, the micro uncertainty will remain, but its impact cannot be empirically determined. Put more strongly, a good data analysis cannot save a bad research design.

To summarize, the hierarchical framework begins with some summary statistics computed at the microlevel and addresses how these summary statistics vary because of macrolevel processes. Within this framework, there are *two* sources of uncertainty, one at the microlevel and one at the macrolevel; both need to be taken into account. In addition, there is important information about the micro-level to be found at the macrolevel. In the illustration, the site mean is *not* the most precise estimate of the site's true snowpack depth. A more precise estimate can be obtained by computing a weighted average of the site mean and the grand mean. But all of these details are moot unless the necessary data are collected.

Extensions to More Realistic Situations

The simple hierarchical framework described above can be extended in a number of ways. First, the causal effect at the microlevel can be made a function of a set of explanatory variables, commonly (but not necessarily) within a linear regression formulation. In effect, δ_j is replaced with the systematic part of some regression equation. Continuing with our illustration, the true snowpack depth can be systematically related to a number of explanatory variables such as when each of the readings were taken. If the readings were spaced over the course of several hours, blowing and/or melting snow could make a difference in the true snowpack depth. Then, δ_j would be replaced by the expression $\alpha_{j0} + \alpha_{j1} x_{ij}$, where the α_j's are the usual regression coefficients, and x_{ij} is time of day. Additional explanatory variables could be represented in a similar fashion.

Second, the impact of the added explanatory variables at the microlevel (i.e., the α_j's) can be made a function of still another set of explanatory variables operating at the macrolevel. In particular, each of the regression coefficients at the microlevel can be made to vary across microlevels because of macro process variables. If, for instance, true snowpack depth at a given site is a function of the time of day when a reading is taken, the impact of the time of day can, in turn, be allowed to vary as a function of whether the measurement rod is in a shaded location. Presumably, the impact of time of day will be less in shaded areas (other things being equal) because the snow around the rod is less exposed to direct sunlight. In other words, one might write $\alpha_{j1} = \beta_0 + \beta_1 z_j$, where the β's are the usual regression coefficients and z_j is the hours of exposure to direct sunlight. A similar equation might be written for α_{j0} and for any other additional microlevel regression coefficients.

Third, one can employ more than two levels in the hierarchy. In our illustration, one might have the set of sites nested within drainage basins. Then there would be three levels: five measures taken at a particular site, a set of sites, and a set of drainage basins.

Fourth, the hierarchical framework can be applied within frequentist traditions, Bayesian traditions, or even a combination of the two. The mathematics is often quite similar; what differs is the interpretation.

Implications and Uses

Perhaps the most important, the hierarchical framework described briefly above provides a formal vehicle for linking natural and social phenomena across different temporal and spatial scales. It is increasingly used with great success in the social sciences (Wong and Mason, 1985) and should be a useful tool in the natural sciences as well.

In addition, uncertainty at one level affects not just the output of the model at that level, but all other levels. In our illustration, haphazard variability in the readings taken at each rod affects estimates of the processes operating across sites that influence snowpack depth. Likewise, haphazard variability in cross-site differences in snowpack depth affects estimates of the processes affecting the readings at reach of the rods. That is, the uncertainty "spills over" (as it should if the phenomena are truly hierarchical).

The uncertainty also has implications for each of the model outputs mentioned earlier. Thus, the uncertainty may be translated into joint distributions for the parameter estimates (e.g., the regression coefficients) at both the microlevel and macrolevel. Marginal distributions, confidence intervals and/or significance tests can follow. The uncertainty may also be translated into joint predictive distributions for forecasts from the model. Marginal distributions and confidence intervals for all forecasts can follow.

Note that even if the parameters in the model are known and fixed, as is often the case in simulations based on some theory, uncertainty in the inputs will have implications for the model's outputs. Suppose, for example, the form of the relationships between precipitation and soil moisture is known precisely. It is still true that uncertainty in the measures of precipitation will translate into uncertainty in forecasts of soil moisture.

Caveats and Conclusions

The hierarchical framework needs to be appreciated as not just an accounting device, but as a representation of a theory. For example, if variation in the microlevel parameters (e.g., regression coefficients) is to be explained by variation in macrolevel variables, the functional form must be a reasonable approximation of the true relationships. Likewise, the hierarchical framework assumes that the microlevel and macrolevel disturbances are uncorrelated. If this assumption is incorrect, the uncertainty will not be captured properly. In short, every aspect of the framework needs to correspond well to the phenomenon being studied.

Another set of issues arises from the data used. In the papers I reviewed, for example, a common problem in mesoscale modeling is working with data from different levels of spatial and temporal resolution. If the level of resolution required is more demanding than the level of resolution available, missing data are the problem. If the level of resolution required is less demanding than the level of resolution available, the problem is how to aggregate.

In the case of missing data, a popular strategy in my readings was to apply some kind of interpolation. However, the interpolated points are *not* really new data, but typically some function of data *already on hand*. So, they cannot be treated simply as additional observations; one cannot just insert the interpolated data and proceed as usual. At the very least, one risks falsely precise estimates because of an artificially inflated sample size (i.e., real data plus interpolated data). The interpolated data also incorporate the uncertainty in the source data used in the interpolation; they are *estimates* of what the missing observations should be. This uncertainty needs to be properly represented in any reported results. In short, missing data create special problems for any analytical proce- dure, including the hierarchical formulation. Proper techniques for handling missing data are available (Little and Rubin, 1987), but they are well beyond the scope of this chapter.

In the case of aggregating over several observations, summary statistics incor- porate the uncertainty of their constituent parts. However, the amount of uncer- tainty in a summary statistic depends on the variability of the data used in the computation and the number of data points. Unless a set of means, for example, is computed from data subsets having the same variability and the same numbers of observations, the means will differ in the amounts of uncertainty they contain. A proper analysis needs to take differential variability into account. In the case of a response variable, for instance, this may mean weighing the data to adjust for differential uncertainty in each observation.

There is another set of problems if any of the coefficients need to be estimated from data (as opposed to being known in advance). All of the usual concerns associated with parameter estimation are relevant. While a discussion is well beyond the scope of this chapter, one illustration will give some sense of the issues.

Considering again the snowpack illustration, suppose that of the several vari- ables that might affect the true snowpack depth at a site, the average number of hours each day exposed to sunlight was a key variable. But also suppose that no direct measure of exposure is available. At the very least, therefore, the model will be incomplete. Equally important, if exposure is related to other variables that are included in the formulation (e.g., elevation), estimates of the coefficients for those variables will improperly capture some of the impact of exposure. That is, confounding will result. Technically, the parameter estimates will be biased, with the size and direction of the bias depending on the size and direction of the associations between exposure and variables included in the model (including reported snowpack depth). In the social sciences at least, such biases can often reverse the signs of estimated effects. There is certainly no mathematical reason why similar problems could not be found in integrated regional models from the physical and biological sciences, although it is also possible that, typically, omitted variables are effectively unrelated to the included variables (i.e., effec- tively just "noise").

To summarize, the hierarchical framework could be a useful tool for taking uncertainty into account in mesoscale models of physical and biological pro- cesses. However, implementing hierarchical models will place heavy demands on a scientist's theory and data. If either the theory or data are seriously suspect, the uncertainty will be inaccurately represented, and misleading conclusions could result. However, weak theory and/or weak data should not be taken as a justification for ignoring uncertainty. Rather, they should underscore the highly tentative nature of any findings and suggest where additional work is needed. Finally, the hierarchical framework is but one way of representing uncertainty that seems especially appropriate for the papers I read. The underlying message is far more general: regardless of what framework is applied, uncertainty would seem to be a scientific fact of life for mesoscale models of physical and biological processes.

References

Barnett, V. 1982. *Comparative Statistical Inference* 2nd ed. John Wiley and Sons, New York.

Bryk, A. S., and S. W. Raudenbush. 1992. *Hierarchical Linear Models*. Sage Publications, Newbury Park, NJ.

Cochran, W. G. 1977. *Sampling Techniques*, John Wiley and Sons, New York.

Cox, D. R., and D. V. Hinkley. 1974. *Theoretical Statistics*. Chapman and Hall, London.

Dickinson, R. E., R. M. Errico, F. Giorgi, and G. T. Bates. 1989. A regional climate model for the western United States. *Climate Change* **15**:383–422.

Draper, D. 1993. *Assessment and Propagation of Model Uncertainty*. Department of Mathmatics, UCLA, Los Angeles, CA.

Giorgi, F., and L. O. Mearns. 1991. Approaches to the simulation of regional climate change—a review. *Reviews of Geophysics* **29**:191–216.

Jacob, D. J., and S. C. Wofsy. 1988. Photochemistry of biogenic emissions over the Amazon forest. *Journal of Geophysical Research* **93**(D2): 1477–1486.

Johnson, N. L. and S. Kotz. 1970. *Continuous Univariate Distributions-1*. John Wiley and Sons, New York.

Little, R. J. A., and D. B. Rubin. 1987. *Statistical Analysis with Missing Data*. John Wiley and Sons, New York.

Maritz, J. S., and T. Lwin. 1989. *Empirical Bayes Methods*. Chapman and Hall, London.

Pielke, R. A., G. A. Dalu, J. S. Snook, T. J. Lee, and T. G. F. Kittel. 1991. Nonlinear influence of mesoscale land use on weather and climate. *Journal of Climate* **4**:1052–1069.

Picard, R. R., and K. N. Berk. 1990. Data splitting. *American Statistician* **44**:140–147.

Picard, R. R., and R. D. Cook. 1984. Cross-validation in regression models. *Journal of the American Statistical Association* **79**:575–583.

Raftery, A. E. 1993. *Approximate Bayes Factors and Accounting for Model Uncertainty in Tenealized Linear Models*. Department of Statistics, University of Washington, Seattle, WA.

Shukla, J., C. Nobre, and P. Sellers. 1990. Amazon deforestation and climate change. *Science* **247**:1322–1325.

Trenberth, K. E., G. W. Branstator, and A. Arkin-Phillip. 1988. Origins of the 1988 North American drought. *Science* **242**:1640.

Urban, D. L., R. V. O'Neill, and H. H. Shugart. 1987. Landscape ecology: A hierarchical perspective can help scientists understand spatial patterns. *Bioscience:***37**:119–127.

Vorosmarty, C. J., B. Moore, A. L. Grace, M. P. Gildea, J. M. Melillo, B. J. Peterson, E. B. Rastetter, and P. A. Steudler. 1989. Continental scale models of water balance and fluvial transport an application to South America. *Global Biogeochemical Cycles* **3**:214–266.

Wong, G. Y., and W. M. Mason. 1985. The hierarchical logistic regression model for multilevel analysis. *Journal of the American Statistical Association:* **80**:513–524.

III

Case Studies

5

Modeling Social Systems and Their Interaction with the Environment: A View from Geography

Diana Liverman

There is a long tradition of attempts to explain the pattern and dynamics of the human use of land and resources. A range of abstractions and computer simulations derived from theory and empirical data—models in the broadest sense—have been used to describe agricultural and industrial use and misuse of the environment.

This chapter provides a brief review of social-science-system modeling as it relates to the social causes and consequences of changes in land use and environmental quality at the regional level. The goal is to explain the assumptions of some key models and their application and to describe some of the limitations of these models and the more general modeling enterprise.

The discipline of geography provides a useful focus for the paper because it has traditionally concerned itself with relationships between humans and the environment at the regional scale and has seen many years of lively debate about the possibilities of modeling human activity (Chorley and Haggett, 1967; Macmillan, 1989). Land use and environmental modeling in geography has drawn from and interacted with many other disciplines such as economics and ecology and has seen a resurgence as a result of the new interest in global change (National Research Council, 1992).

The Question of Regions

With regard to the question of *regional* modeling, geographers have discussed and debated the concept of the region for many years (Gilbert, 1988; Johnston and Gregory, 1986; Hartshorne, 1939). One view is that regions are areas of essentially uniform physical or social characteristics. The classic definition of a region is to use physiographic or other "natural" features to divide the earth into relatively homogeneous areas based on topographic, climatic or vegetation characteristics. In this uniform approach to regionalization, boundaries are drawn where there are distinct shifts in topography or rainfall. Uniform regions may also be based on socioeconomic characteristics. In contrast, the functional or nodal method defines regions with reference to a major city or other center of influence and interaction.

Most regionalizations are developed according to *a priori* or *deductive* criteria, in which the classification of units into regions of common characteristics is achieved through specifying levels of income, language or social conditions, which are assumed to divide the earth and nations into regions. Alternative methods of regionalization use statistical procedures in an *inductive* approach to classify data, from small physical or social units, into regions in which common features are identified in the statistical characteristics of the units. For example, units might be clustered according to their industrial structure or demographic characteristics using a principal-components analysis of census data.

Perhaps the most important lesson of the debate about regions in geography is that both the definition of regions and the regions themselves are very dynamic. There are no immutable or sharp boundaries on the earth's surface (especially in the social realm) and in many cases research on regions must take as its first major task the identification and bounding of the region to be studied. In interdisciplinary, integrated regional studies, one of the challenges is to reconcile the approaches of disciplines using different data at different scales; for example, those of social scientists who work with administrative data at the scale of states and nations, and the environmental scientist who uses river basins or ecosystems to define the region.

In many cases, it is reasonably easy to arrive at a common definition of a region because they often seem to emerge naturally from the long historical development of human interactions with landscapes. In the United States, such regions include the Great Plains, the Midwest, the Great Lakes, the Los Angeles Basin, or New England. Geographic Information Systems provide the opportunity to overlay a variety of regional maps in order to identify appropriate commonalities and boundaries for regional analysis.

Environmental or Physical Models of Land Use

One important set of models explains land use patterns and dynamics in terms of physical and biological characteristics of the environment or region. For example, there are a range of models that have been used to estimate the best or most probable use of agricultural land based on climate, soils and topography. The work of geographers such as O.E. Baker (1926), who developed a regionalization of American agriculture based primarily on physical factors, and Griffith Taylor (1930), who identified the environmental constraints on land use and human settlement in Australia, provide early descriptive examples of this approach.

Similar assumptions about environmental controls on land use emerge from a number of recent and more quantitative studies on the agricultural potential of different regions and the world as a whole. Clark (1967) used climatic data to calculate a world area of productive land of 7.7–10.7 billion hectares depending on how one calculates the productivity of tropical soils. He estimated that this land could support up to 49 billion people. A similar study by Revelle (1976) estimated enough land to feed 40 billion people. In a detailed study Buringh (1977) superimposed climate and soil maps to identify regions of arable and grazing potential. They concluded that 24% of global land (3.2 billion hectares) had potential for crop production. Large areas of potentially arable, but currently uncultivated, land were identified in South America, Africa and Australasia.

The work of Buringh and colleagues was the basis for a major study of the food production potential of lands in the developing world (Linnemann et al., 1979). The potential production of hundreds of soil–climate regions was estimated using simple models for 16 major crops and assumed that the highest yielding crop would be planted on each unit (Harrison, 1983). The study allowed for soil fertility and moisture constraints and for three levels of technological inputs (e.g., fertilizer or irrigation). The study indicated that Africa is only using 21% of its potential arable area and Latin America only 11% but that Southeast Asia is nearing its land limits at 92%. If the potential food production is compared to current population levels, these models suggest that a number of regions are viewed as exceeding the capability of the land to support people, i.e., their carrying capacity.

Agricultural potential models that assume that technology can be used to expand production often fail to account for the availability of capital to purchase the technology (e.g., many regions are too poor), or for the ways in which negative environmental impacts of technology—such as erosion, pollution and plant breeding—can feed back and destroy the resource base of soils, water, and genetic diversity.

A parallel set of studies in ecology explain patterns of global vegetation and biomass production primarily in terms of climate (Lieth and Whittaker, 1975). The major controls on the dominant species in each region are assumed to be variables such as temperature, rainfall, and soil type. Other factors such as competition or human modification are not considered.

The dynamic version of these agricultural and ecological models can be seen in the escalating number of studies that assess the possible impacts of climate change on crop yields and vegetation (Bolin et al., 1988). Adams et al. (1990) use the output of global climate models to estimate how crop yields may change in the United States as a result of global warming. Parry et al. (1988) use a similar methodology to model the impacts of both past climate variability and possible future scenarios on crop production in different regions of the world.

Emmanuel et al., (1985) have used climate model output to perturb empirical models of the relationship among climate, vegetation type, and vegetation productivity. These climate-change studies tend to assume that only climate (and often only average temperature and precipitation) will change; that climate is the most important influence; and that other factors—such as soil or technology—are assumed not to alter.

Most of the models discussed above are empirically based. The simplest are regression models of the form $y = ax + b$, where y might be crop yield or biomass and x is temperature. The use of more physiologically or physically based models, which simulate processes such as photosynthesis, evaporation and nutrient flow, is limited at the regional scale by the data needs and complexity of the models. Thus, the CERES model used by Adams et al., (1990) to assess climate impacts on maize and wheat yields is heavily parameterized in simulating the day to day effects of water and nutrient stress on plant growth.

These models provide an explanation of land use grounded in a tradition of geography called environmental determinism, wherein the physical environment, particularly climate, is seen as the most important influence on human activity. The more extreme versions of environmental determinism are associated with the work of Ellen Churchill Semple (1911) and Ellsworth Huntington (1962) who presented physical geography as the major explanation of not only land use but levels of development, culture, and "civilization." In this framework, regions are poor because the physical environment has limited economic growth and human potential. Although there was a strong reaction against both the scientific accuracy and political implications of environmental determinism (Peet, 1985) because of its oversimplification and links to racism and imperialism, environmental explanations of land use patterns and dynamics are still prevalent, and, many would argue, relevant.

Environmental explanations of land use and land use change might also include models that treat humans as a purely biological organism whose demography responds to and has impact on the environment. Influenced by the views of Darwin, Thomas Malthus developed a model of the relationship between agricultural production and human population growth that predicted famine as a rapidly growing population outgrew a fixed or slowly growing food supply.

The Malthusian model has become important in many recent discussions of land use and degradation, where the intensification of land use and the accompanying problems of erosion, deforestation and pollution are associated with rapid population growth. The work of biologists Paul Ehrlich and Garrett Hardin is characteristic in linking agricultural production and population through the concept of carrying capacity: the ability of the land to support a population based on physical resources, average per capita food needs, and population size (Ehrlich and Ehrlich, 1990; Hardin, 1972). If the population exceeds carrying capacity, the model suggests that hunger, environmental degradation and the destruction of biodiversity will ensue. This model also explains the rapid conversion of natural ecosystems to agricultural land to feed a voracious and fast breeding, dominant human organism at the top of the food chain.

Economic and Social Explanations of Land Use

A different approach to modeling is based in economic explanations of land use. The classic model is that which was developed by Von Thünen (1966) that attempts to explain the agricultural use of land on a uniform plain around a single, isolated market. In Von Thünen's model, land uses are determined by the cost of transport to market, which depends on the distance to market and the bulk and perishability of commodities. The value of each crop to farmers (i.e., the surplus profit after production and transport costs have been paid—also called the economic rent) declines with distance from the market such that bulky or perishable products in high demand are produced close to the market, and durable, less profitable products are produced farther away. This results in concentric zones of agricultural land use around a city. In Von Thünen's time, dairy and vegetables were grown in the closest ring, then the wood needed for fuel and construction, and then grains in a series of zones based on crop rotations (Fig. 1).

Scholars have relaxed the assumptions of the simple Von Thünen model, allowing for variations in environmental conditions, transport infrastructure and technology, or number of markets. The modified model has been used to explain land use patterns in places as varied as Uruguay, Brazil, and the northeast United States (Kolass and Nystuen, 1974).

Similar economic assumptions about transport costs and markets also emerge in the industrial location models of Weber and Losch, and the settlement pattern model of Christaller. Weber assumed that industry will locate so as to minimize costs of labor and raw material transport and to take advantage of industrial clusters or agglomerations (Chisholm, 1966; Haggett, 1969). Losch focused on access to markets and the minimization of costs. Christaller described a theoretical model of settlement in which a hierarchy of central places providing services develops in polygonal patterns on a uniform plain. The most important assump-

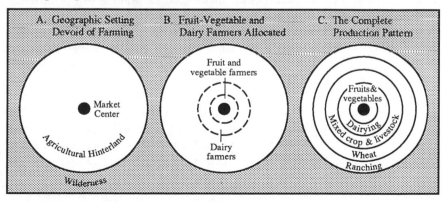

Figure 1. Von Thünen's isolated state model in its stages of formation. (Source: Liverman, 1986.)

tions of these economic land use models include those of cost (especially transport) minimization and profit maximization in free markets with rational, optimizing producers with complete and instantaneous information about prices and costs. These models are typically associated with the neoclassical tradition in economics.

One major criticism of such models focuses on their assumptions of optimum behavior and perfect information. Geographers and other social scientists have developed behavioral models that assume less rational use of land, taking into account factors such as culture, inaccurate information, and suboptimal behavior. Wolpert's (1964) study of farm decision-making in Middle Sweden assumed that farmers decisions about land use were constrained by factors such as information, cultural and educational background, and that decisions would be "satisfying" rather than optimum. The availability of information could be simulated using diffusion models of the flow of information and technology taking into account the influence of space and physical and cultural communication barriers (Hagerstrand, 1968). Those farthest from centers of innovation receive information later and may use the land in more traditional ways.

Diffusion models have also been applied to the widespread transformations in land use associated with the Green Revolution, which brought the technologies of plant breeding and chemicals to agriculture in the Third World (Yapa, 1977).

This type of land use model often uses probabilistic Monte-Carlo-type methods to illustrate some of the uncertainty and process in human decisions. Agricultural and other economists have also developed much more complex models of land use and environmental impacts in which some, but not all, of the neoclassical assumptions are relaxed. For example, linear programming models permit the optimization of locational and transport decisions based on a range of economic and physical constraints.

A more descriptive approach to land use decisions is associated with the cultural ecology of Carl Sauer which focused on the ways in which different cultures adapt to and transform their environment (Denevan, 1989). Cultural ecologists have documented the variety of ways in which traditional peoples adapted to constraints of climate and terrain through terracing, raised fields and irrigation and thus transformed severe landscapes such as the Andes and Amazon forest to agricultural land (Browder, 1989).

Another view comes from political economy. In this case, the human use of land is constrained not by personal characteristics or information diffusion but by structures of economic and political power. Thus the agricultural history and geography of the United States is explained less by environment and the free market economy and more by factors such as cheap labor (e.g., slavery and the exploitation of farmworkers), the concentration of land and speculation in its value, and the intervention of government in the interest of certain powerful lobbies such as the railroad barons and agribusiness (FitzSimmons 1987; Goodman and Redclift, 1991). Third World land use and environmental degradation is explained through the legacy of colonialism in export-oriented economies and land concentration, and the unequal participation of regions in the international economy (Myrdal, 1971; Smith, 1984).

Integrated Models

The sharp distinction between environmental and socioeconomic models of land use is blurred in many studies as authors relax the assumptions of their models. Incorporating variations in soil fertility in the Von Thünen model or technology in potential production studies provides more accurate reproduction of actual land use patterns. For example, a basic geography text by Kolars and Nystuen (1974) shows how the land use of the United States can be replicated in a modified Von Thünen model (Fig. 2).

One set of integrated approaches is empirical models based on statistical relationships. In many cases these models include both environmental and economic explanatory variables to estimate crop yield, crop area, or measures of environmental degradation. The Universal Soil Loss Equation, frequently used to assess soil erosion potential, uses both environmental (slope, rainfall) and social (farm practices) variables (Larson et al., 1983). The degree to which variables are preselected based on theoretical grounds for use in the models varies. Allen and Barnes (1985) use a large number of variables including population growth, income, export and land use measures in their attempt to estimate deforestation rates in developing countries. Such correlation and regression mod-

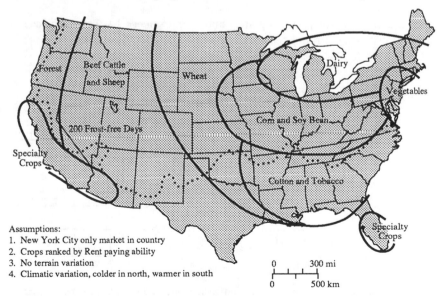

Figure 2. Theoretical Land use rings in the United States. (Kolars and Nystuen, 1974).

els select variables on statistical rather than theoretical criteria. Econometric and linear-programming models such as those of Leontiev et al. (1977) and Heady and Egbert (1964) relate land use and agricultural production to changes in input availability and demand. In these models, environmental conditions may act to influence input needs for fertilizer and irrigation, or as constraints on land expansion.

Recent developments in environmental economics also link economic and environmental variables in the integration of environmental values into input–output models and cost-benefit analyses. Those models of regional economies that link production changes to both economic and environmental impacts are particularly useful in regional economic modeling.

The most complex and ambitious integrated models of the human–environment relation are world simulation models of which the best known is the World 3 model used by Meadows et al., in the book *Limits to Growth* (1972). This model, which treated the world as one region, projected rapid increases in arable land use until about 2020 when costs, urbanization, and erosion begin to reduce the amount of land in production. Figure 3 shows how one world model—the International Future's Simulation (IFS)—structures the links between four sectors and ten regions and responds to land resource constraints and government policy. IFS has been used to assess the possible impacts of global climate change on the world food system through changes in crop yields and area (Liverman, 1989).

The Limitations of Modeling

The modeling enterprise in geography and social science has been subject to a range of criticisms. From a technical standpoint, models have been criticized as

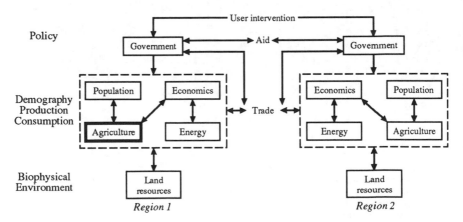

Figure 3.   International future's simulation basic model framework. (Source: Liverman, 1986.)

too simplistic, too aggregate, oversensitive or unable to reproduce real world conditions (Cole et al., 1973; Liverman, 1989). More fundamentally, human geography and social science have evolved substantive ideological and epistemological condemnations of the formal, especially quantitative, approach to explaining society. In geography, the reaction against modeling and quantitative social science was mainly from those with humanistic and political economy perspectives (Johnston, 1980; Pepper, 1984). Humanists objected to the generalizations regarding individual behavior and the quantification of intangible values in economic models. Political economists felt that the models did not capture the power relations in the political economic structure, ignored the role of the state, and were themselves a tool of elite, technocratic managers (Cole et al., 1973; Deutsch et al., 1977). Recent objections to the modeling of nature–society relations include those of the deep ecologists who object to the anthropocentric separation of nature and society; realists who suggest that models are abstractions and cannot reproduce or forecast particular contingent conditions; feminists who see models as patriarchal or inattentive to the particular conditions of women; and postmodernists who argue that there are many different interpretations and meanings of any one phenomenon.

These multiple criticisms mean that social scientists have become rather self-conscious and self-critical in attempts to model society and human–environment interactions. Quantitative modeling may be viewed as technically impossible, socially meaningless, or even politically dangerous by colleagues.

On the other hand, in a world where regions and economic sectors are increasingly interdependent, or where the understanding of environmental relations becomes ever more detailed, models offer us the possibility of managing large amounts of complex information and relationships. They force us to specify very explicitly what we know and do not know in a form that should, in some ways, be very transparent (if the model is well documented) and can thus be evaluated and modified by others, even if they speak a different language. Simulation models offer the possibility of formalizing assumptions, organizing relevant data, and specifying links and feedbacks between different economic sectors and geographical regions. The collection of data for constructing and validating a model can reveal inadequacies in the empirical evidence about environment and land use, and in a similar way, the process of constructing equations to describe the system can show where theory is inadequate. Finally, models allow us to undertake experiments and test out policy options that would be impossible or ethically unacceptable as untested experiments in the real world.

References

Adams, R. M., C. Rosenzweig, R. M. Peart, J. T. Ritchie, B. A. McCarl, J. D. Glyer, R. B. Curry, J. W. Jones, K. J. Boote, and L. H. Allen, Jr. 1990. Global climate change and US agriculture. *Nature* **345**:219–224.

Allen, J. C., and D. F. Barnes. 1985. The causes of deforestation in developing countries. *Annals of the Association of American Geographers*. **75**:163–184.

Baker, O. E. 1926. Agricultural regions of North America. *Economic Geography* **2**:459–493.

Barney, G. O. 1980. *The Global 2000 Report*. Pergamon, New York.

Bolin, B. (ed). 1986. *The Greenhouse Effect, Climate Change and Ecosystems. SCOPE 29*. John Wiley and Sons, New York.

Browder, J. (ed.). 1989. *The Fragile Lands of Latin America*. Westview, Boulder, CO.

Buringh, P. 1977. Food production potential of world. *World Development* **5**:477–485.

Chisholm, M. 1966. *Rural Settlement and Land Use*. Humanities Press, Atlantic Highlands, NJ.

Chorley, R. J., and P. Haggett (eds.). 1967. *Models in Geography*. Methuen, London.

Clark, C. 1967. *Population Growth and Land Use*. St. Martin's, New York.

Clark, J., and S. Cole. 1975. *Global Simulation Models: A Comparative Study*. Wiley, London.

Cole, H. S. D., C. Freeman, M. Jahoda, and R. L. R. Pavitt (eds.). 1973. *Models of Doom*. Universe Books, New York.

Denevan, W. M. 1989. Chapter 1. In J. Browder (ed.). *The Fragile Lands of Latin America*. Westview, Boulder, CO.

Deutsch, K. W., B. Fitch, H. Jaguaribe, and A. Markovitz (eds.). 1977. *Problems of World Modeling*. Ballinger, Cambridge, MA.

Ehrlich, P., and A. Ehrlich. 1990. *The Population Explosion*. Simon & Schuster, New York.

Emmanuel, W. R., H. H. Shugart, and M. P. Stevenson. 1985. Climate change and the broad scale distribution of terrestrial ecosystem complexes. *Climatic Change* **7**:29–43.

FitzSimmons, M. I. 1987. The new industrial agriculture. *Economic Geography* **62**:334–353.

Gilbert, A. 1988. The new regional geography in English and French-speaking countries. *Progress in Human Geography* **12**:208–228.

Goodman, D., and M. Redcliff. 1991. *Refashioning Nature*. Routledge, New York.

Hagerstrand, T. 1968. *Innovation Diffusion as a Spatial Process*. Univ. of Chicago Press, Chicago, IL.

Hagerstrand, T. 1952. The propagation of innovation waves. *Lund Studies in Geography. Series B. Human Geography* **4**:3–9.

Haggett, P. 1969. *Locational Analysis in Human Geography*. St. Martin, New York.

Hardin, G. 1972. *Exploring New Ethics for Survival: The Voyage of Spaceship Beagle*. Viking, New York.

Harrison, P. 1983. Land and people, the growing pressure, in FAO Economic and Social Development Series, *Land, Food and People*. FAO, Rome.

Hartshorne, R. J. 1939. *The Nature of Geography*. Association of American Geographers, Lancaster, PA.

Harvey, D. 1966. Theoretical concepts and the analysis of agricultural land use patterns in geography. *Annals of the Association of American Geographers* **56**:361–374.

Heady, E. O., and A. C. Egbert. 1964. Regional programming of efficient agricultural production patterns. *Econometrica* **32**:374–386.

Hughes, B. B. 1981. *World Modeling*. Lexington, Lexington.

Huntington, E. 1962. "The distribution of civilization. In E. Huntington (ed.). *Mainsprings of Civilization*. Ayer Co. Publishers, Salem, NH, pp. 391–408.

Johnston, R. J. 1980. *Geography and Geographers*. Halsted Press, New York.

Johnston, R. J. and D. Gregory. 1986. *The Dictionary of Human Geography*. Basil Blackwell, Oxford.

Kolars, J. F., and J. D. Nystuen. 1974. *Human Geography: Spatial Design in a World Society*. McGraw Hill, New York.

Larson, W. E., F. H. Pierce, and R. H. Dowdy. 1983. The threat of soil erosion to long term crop production. *Science* **219**:458–465.

Leontief, W., A. Carter, and P. Petrie. 1977. *Future of the World Economy*. Oxford University Press, New York.

Lieth, H., and R. H. Whittaker. 1975. *The Primary Production of the Biosphere. Ecological Studies No. 14*. Springer-Verlag, New York.

Linnemann, H., J. de Hoogh, M. A. Keyzer, and H.D.J. Van Heemst. 1979. *MOIRA: Model of International Relations in Agriculture*. North Holland, Amsterdam.

Liverman, D. M. 1986. The response of a global food model to climate change and variation: A sensitivity analysis of IFS. *Journal of Climatology* **6**:355–373.

Liverman, D. M., and K. Frohberg. 1988. Preliminary results of climate change scenarios in the IIASA-FAP model (unpublished report).

Liverman, D. M. 1989. Evaluating global models. *Journal of Environmental Management* **29**:215–235.

Losch, A. 1954. *The Economics of Location* Yale University Press, New Haven, CT.

Macmillan, B. (ed.). 1989. *Remodelling Geography*. Basil Blackwell, London.

Meadows, D. II., D. L. Meadows, J. Randers, and W. K. Behrens III. 1972. *Limits to Growth*. Universe Books, New York.

Meadows, D. H., W. Richardson, and G. Bruckman. 1983. *Groping in the Dark: A History of the First Decade of Global Modeling*. John Wiley and Sons, New York.

Mesarovic, M., and E. Pestel. 1974. *Mankind at the Turning Point*. E. Dutton, New York.

Myrdal, G. 1971. *Economic Theory and Underdeveloped Regions*. Harper & Row, NY.

National Research Council. 1992. *Global Environmental Change: The Human Dimensions*. National Academy of Sciences, Washington, DC.

Office of Technology Assessment (OTA). 1982. *Global Models, World Futures and Public Policy: A Critique*. OTA, Washington, DC.

Parry, M. L., T. R. Carter, and N. T. Konjin (eds.). 1988. *The Impact of Climate Variations on Agriculture*. Reidel, Dordecht, Holland, Vols. 1 and 2.

Peet, R. 1985. The social origins of environmental determinism. *Annals of the Association of American Geographers* **75**:309–333.

Pepper, D. 1984. *The Roots of Modern Environmentalism*. Chapman & Hall, London.

Revelle, R. 1976. The resources available for agriculture. In *Scientific American's Food and Agriculture*. W.H. Freeman, San Francisco, CA, pp. 113–125.

Robinson, J. 1985. Global modeling and simulation. In R.W. Kates, J. Ausubel, and M. Berberain (eds.). *Climate Impact Assessment*. John Wiley and Sons, New York.

Semple, E. C. 1911. *Influences of Geographic Environment*. Holt and Company, New York.

Smith, N. 1984. *Uneven Development: Nature, Capital and the Production of Space*. Blackwell, London.

Taylor, G. 1930. Agricultural regions of Australia. *Economic Geography* **6**:109–134.

Thomas, R. W., and R. J. Huggett. 1980. *Modelling in Geography*. Barnes and Noble, Totowa, NJ.

Von Thünen, J. H. 1966. *Isolated State*. (an English translation of *Der Isolierte Staat*. Translated by Carlam Wartenberg) Pergamon Press, New York.

Wolpert, J. 1964. The decision process in a spatial context. *Annals of the Association of American Geographer* **54**:537–558.

Yapa, L. S. 1977. The Green Revolution: a diffusion model. *Annals of the Association of American Geographers* **67**:350–359.

6

Interactions of Landuse and Ecosystem Structure and Function: A Case Study in the Central Great Plains

*Ingrid C. Burke, William K. Lauenroth,
William J. Parton, and C. Vernon Cole*

Introduction

Spatial pattern in ecological phenomena has been an important impetus for ideas about controls over ecosystem processes. Many of the significant theories about ecosystem function have been based on observation of landscape-scale pattern (Bormann and Likens, 1979; Watt, 1947), chrono-sequences (Whittaker, 1953; 1973), or geographic scale pattern (Burke et al., 1989; Jenny, 1930; McArthur, 1972; Sala et al., 1988). Such observations play two important roles. First, generalizations about large-scale, steady-state patterns provide hypotheses about mechanisms that control ecosystem function which can be tested at small scales and over short time intervals and can be used to generate mechanistic models. Second, such observations provide mathematical/statistical relationships that can be applied to predict future conditions under scenarios in which the controls change. An example is the Holdridge approach to predicting vegetation using climate-change scenarios (Emanuel et al., 1985; Holdridge, 1947).

Both natural history and social sciences have a significant basis in geographic analysis and in studies that attempt to identify the mechanisms responsible for spatial pattern. Much human geography has focused on the correspondence among patterns in populations, cultural predispositions, and natural resource geography (Burton et al., 1978; Lowenthal, 1961; 1972). Such work has stimulated field studies at local levels on the dependence of human cultures on environment. In these studies, the natural environment is treated as the set of independent variables and the human environment as the dependent variables.

We now recognize that long-term and large-scale environmental change and human welfare are closely linked, and that predictions of future conditions must be made with explicit consideration of these linkages (Riebsame et al., 1994; Stern et al., 1992). Such analysis necessitates new kinds of interdisciplinary research. Spatial analysis of broadscale patterns is an appropriate and useful way to generate ideas about specific linkages.

In this chapter, we review two of our recent studies in the central Great Plains of the United States. The objective of this work has been to evaluate two of the key sets of human–environment interactions that occur in this region. First, we review a simulation study that illustrates the influence of land use management on regional patterns of ecosystem structure and function, and the effects of these interactions on the atmosphere. Second, we review a statistical study that had the purpose of assessing the relationship between physical environmental factors and extant land use patterns for the region, and of drawing conclusions regarding the potential importance of socioeconomic factors in controlling land use. We chose the central Great Plains as a study area, a region comprising Kansas, Nebraska, and eastern Colorado, because many of the major land use types of the central Grassland Region occur within this area and because of the significant environmental and land use gradient that occurs across the area.

The conceptual model for our work in the central Great Plains includes the idea that there are significant, regional-scale interactions among climate, soils, land use, ecosystem structure and function, and socioeconomic factors (Fig. 1). For example, Jenny (1980) suggested that climate, parent material, time, and topographic location were major determinants of ecosystem properties. At re-gional scales in the central Great Plains, time and topography can be considered

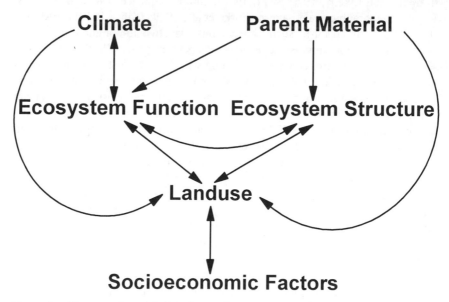

Figure 1. Conceptual model of the interactions among climate, ecosystem structure and function, land use, and socioeconomic factors that are important at a regional scale in the grasslands region of the United States.

to be relatively constant. Thus, climate and parent material are the major natural determinants of ecosystem structure and function, influencing species and life-form distributions, net primary production, and soil organic matter dynamics. Land use, however, is also an important control over ecosystem properties, as is abundantly clear from the very large areas of cultivated land across the region. The major effects of land use are to alter the dominant plant species, to reduce biological diversity, and to change the dynamics of soil organic matter, and potentially to have important feedbacks to atmospheric processes through gaseous, radiative, and hydrologic interactions (Burke et al., 1991; Elliott and Cole, 1989; Pielke and Avissar, 1990). Because particular land use practices are only feasible under certain conditions, variables such as climate, soils, and ecosystem structure significantly constrain land use patterns. Finally, it is clear that regional, national, and global socioeconomic conditions can play an important role in influencing land use decisions (Meyer and Turner, 1992; Riebsame et al., 1994). The interactions and linkages among these system elements are not well understood at either large spatial or long temporal scales.

Regional Data

To assess relationships among physical environmental variables, ecosystem behavior, and land use patterns, we have obtained regional data on climate, soils, and land use for the central Great Plains. These data were entered into a geographic information system (GIS) to allow us to use them as input variables in a simulation analysis and for multivariate statistical analysis. Below, we briefly describe each of the regional data sets.

Climate data were obtained from the CLIMATEDATA database (CLIMATEDATA, 1988) for more than 400 U.S. Weather Bureau stations within the region. Mean annual precipitation and mean annual temperature were interpolated to generalize the information spatially using procedures described in Burke et al. (1991). The climate pattern in the region is typical of that of the central Grassland Region of the United States (Borchert, 1950). Precipitation increases in an easterly direction, and temperature increases in a southerly direction (Fig. 2a and b).

Soils data were obtained from the Soil Conservation Service STATSGO database of soil associations (USDA Soil Conservation Service, 1989). Because the ecosystem simulation model requires input data on soil texture only, we estimated sand and clay content from map unit descriptions of soil textural class (Burke et al., 1991) (Fig. 3). Much of the central Great Plains is characterized by eolian soils, with particularly sandy soils located in eolian basins from the late Holocene, such as the Sand Hills of Nebraska and along the South Platte and Arkansas Rivers (Muhs, 1985). Shallow, fine-textured soils are located in the Flint Hills of Kansas, on soils that developed from shale. Minimum spatial resolution of

Mean Annual Precipitation (mm)

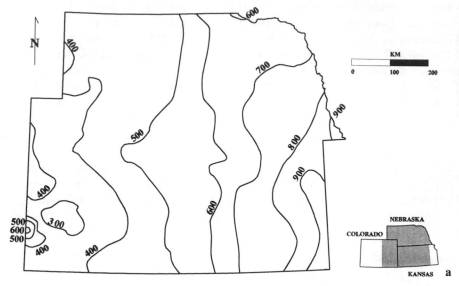

Mean Annual Temperature (C)

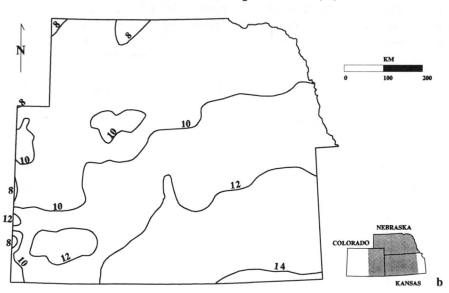

Figure 2. Contoured mean annual precipitation and temperature in the Central Great Plains of the United States. Data were generalized from point data (CLIMATEDATA, 1988).

these data is approximately 1 km^2. State boundaries evident in the data indicate problems with data quality.

Land use data were obtained from two different sources. Remotely sensed imagery provided by the USGS Eros DATA Center, Sioux Falls, SD (Loveland et al., 1991), allowed us to subdivide the region into crop types (Fig. 4), which was needed for the simulation analysis. These data were generated from the Advanced Very High Resolution Radiometer data (1 km^2 resolution), utilizing an initial spectral classification that was corrected using climate information. For this reason, a second land use database was necessary for statistical comparison of land use patterns with climate and soil data. We used the USGS land use Land Cover Data (USDI, U.S. Geologic Survey, 1990), providing land use information at Anderson Level I (Anderson et al., 1976) with a 200 m^2 resolution. Anderson Level I is the coarsest-scale classification available. The information we used was primarily for cropland and rangeland.

The land use data show strong regional patterns. Rangeland predominates in the driest western part of the region, with increasing cropland area in an easterly direction. Dryland wheat/fallow systems occur in the middle of the region, and continuous wheat and dryland corn occur in the far eastern section. Irrigated corn occurs near areas of significant surface or subsurface water. We will describe the reasons for these patterns in the section on Environmental Constraints on land use.

Influence of Land Use on Ecosystem Structure and Function

A large number of studies have been conducted on local-scale effects of cropland management on ecosystem structure and function (Doran, 1980; Haas et al., 1957; Hide and Metzger, 1939; Holland and Coleman, 1987). Most regional work to date on land use–ecosystem interactions has focused on the tropics and the largest number of these have investigated effects of forest clearing (Houghton, 1990; Houghton et al., 1987; Post et al., 1990).

The objective for our study was to estimate the long-term, large-scale impacts of cropland management on soil organic carbon levels in the central Great Plains (Burke et al., unpublished). It has long been recognized that cultivation practices significantly influence ecosystem structure and processes. A good long-term integrator of these processes that has regional relevance is soil carbon. Soil carbon represents the long-term balance of productivity, decomposition, and erosion, and in semiarid regions is the single best indicator of ecosystem stability and sustainability (Burke et al., 1989). In addition, because of its interaction with global atmospheric carbon pools, it is important to large-scale "global change" studies (Schlesinger, 1990). We evaluated the long-term, historical effects of cropland management on soil organic carbon on the central Great Plains.

We used the CENTURY model (Parton et al., 1987) to assess soil carbon losses as a result of fifty years of historical cultivation practices. The CENTURY model is an ecosystem model developed for grasslands that simulates the inputs, outputs, and turnover of soil organic matter as a three-compartment system. The model has been validated extensively for grasslands and cultivated lands within the Central Grasslands region of the United States (Burke et al., 1989; Parton et al., 1987). We assumed that land management patterns across the region were constant during the fifty year period from 1900 to 1950. We used the GIS to organize the major driving variables for the model, including climate, soil texture, and land use. We overlaid the maps in the GIS and conducted a simulation for each unique combination of input variables, running the model to steady state to estimate native soil carbon across the region (Fig. 5). Annual variation in temperature and precipitation was randomly generated for climate classes based on information on the mean monthly data and their variance (Burke et al., 1991). The model parameters were changed for each cropland management practice according to historical information on crop varieties, tillage intensity, and residue return rates, and validation was conducted by Metherill (1992). Mean simulated grain production for each crop type was compared to historical grain yields from records for Weld County, CO, and state records for Kansas (Kansas State Board of Agriculture, 1990). Details about model parameters, the GIS linkage to the simulation model, and explanations for model behavior may be found in Burke et al. (unpublished). Rangelands were represented with moderate levels of grazing (30% removal of aboveground biomass.)

The results of the simulations indicated that 35–50% of soil organic matter was lost as a result of fifty years of cultivation in the region (Fig. 6). Simulated losses occur entirely as a result of decomposition, and do not include erosional losses. These results correspond well with a large number of published studies (Aguilar et al., 1988; Burke et al., 1989; Haas et al. 1957). Cultivation causes soil organic matter losses as a result of decreased inputs of above- and below-ground plant material, as well as increases in decomposition rate owing to disruption of soil aggregates and increased residue contact with decomposers in the soil (Aguilar et al., 1988; Tiessen et al., 1982). Our simulations suggested all of the croplands had significant losses of soil organic matter but that rangelands did not. Dryland wheat–fallow systems had the highest rate of soil organic matter losses, owing to the long fallow period with intensive tillage and no plant production (Fig. 7). Intermediate losses occurred in irrigated corn because of the relatively high tillage intensity, and lowest simulated losses occurred in continuous wheat and dryland corn. Significant interactions with precipitation, temperature, and texture occurred, with the general pattern that areas with the highest amounts of native soil carbon lost the largest amounts of soil carbon.

Our results suggest that historical cultivation over much of the central Great Plains has had significant impacts on regional ecological properties. Across all cultivated areas in the region, an average of 44% of soil carbon in the surface

Sand Content

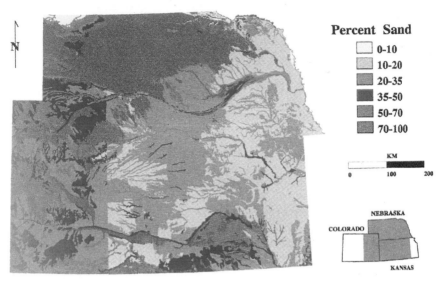

Figure 3. Sand content of soils in the Central Great Plains of the United States. Data were generalized from soil association maps and attribute data (USDA Soil Conservation Service, 1989).

Agricultural Land Use

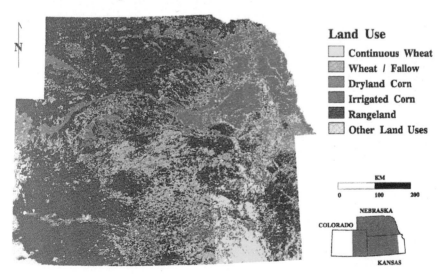

Figure 4. Land use in the Central Great Plains of the United States. Land use classes were determined from special data from the Advanced Very High Resolution Radiometer (Loveland et al., 1991).

Soil Carbon - Steady State 1900

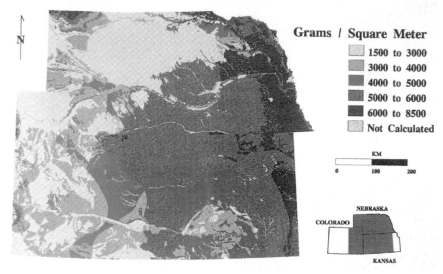

Figure 5. Simulated steady-state soil organic carbon. Results of long-term simulations using the CENTURY ecosystem model (Parton el al., 1987) with geographic inputs on climate and soil texture.

Soil Carbon Change 1900-1950

Figure 6. Simulated soil carbon losses due to fifty years of historical cultivation. The map was generated by applying the CENTURY model (Metherill, 1992; Parton et al., 1987) to geographic data on climate, soil texture, and land use.

Soil Organic Carbon

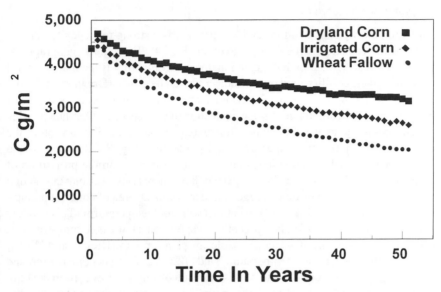

Figure 7. Simulated losses of soil carbon due to cropping using wheat–fallow, dryland corn, and irrigated corn systems.

20 cm was lost; across the entire region (i.e., all land use types), 21.5% of surface carbon was estimated to be lost from fifty years of cultivation. Soil organic carbon is a fundamental characteristic of ecosystems, representing long-term system fertility and sustainability. Ecosystem carbon is predominantly located in soils of semiarid systems, and the simulated losses as a result of cultivation represent a substantial potential contribution to atmospheric CO_2. Although improvements in cropland management can increase stored carbon (Burke et al., unpublished), such increases occur at pedogenic rather than human time scales (decades) (Schlesinger, 1990). Thus, long-term land use management may have significant and persistent regional implications.

Environmental Constraints on Land use

What controls land use pattern in the central Great Plains? Observations suggest that long-term patterns are strongly related to environmental factors. In this section, we will first present an assessment of land use and crop-type patterns across this region, based on observations of geographic patterns. We will next present a statistical analysis of the degree to which environmental constraints can explain the variance in patterns of cropland and rangeland. Finally, we will suggest additional, socioeconomic explanations for land use patterns and present suggestions for future research.

Cropland vs Rangeland

An important regional trend in land use patterns is a strong tendency toward an increase in cropland with easterly direction (Fig. 4). Ecologists, geographers, and agronomists have long recognized that water availability is the major control over cropland feasibility, and over which crops can be grown (Dregne and Willis, 1983). These observations are based on large-scale qualitative assessments of geographic pattern and on field-scale evaluation of crop success. Our data provide the opportunity to assess the relationship among cropland distribution across the region and precipitation, temperature, and soil texture (Fig. 8). Between 30 and 60 cm mean annual precipitation, there is a strong increase in the proportion of cropland area, relative to total land area (Fig. 8a). There is also a strong interaction with mean annual temperature, suggesting that within an area of constant precipitation cropping is more feasible at locations with cool temperatures. This is likely due to the influence of higher potential evaporation at warmer temperatures, decreasing the amount of water available for plant growth (Dregne and Willis, 1983). Within the warmest temperature band (12°C), a strong decline in cropland is evident in our data above 70 cm; this is the result of very shallow soils in the Flint Hills of Kansas that happen to occur spatially within this particular temperature interval.

The second major trend in the data is a strong correspondence in the distribution of cropland with soil texture (Fig. 8b). The land use and soil maps indicate that in the areas that are marginal for cropping (roughly 50% of the area is cropped), the presence of very sandy soils correspond with rangelands. This occurs in areas that fall south of the Platte River but are not irrigated, and in the Sand Hills of Nebraska. It is well recognized that sandy soils have a low water-holding capacity and high erodability, which limits their utility as cropland soils. In addition, sandy soils have a low organic matter content owing to low soil aggregation and protection of soil organic matter from decomposition (Jenkinson, 1977; Paul 1984; Sorenson, 1981), thus, lower nutrient supply for plant growth. The only areas within the region that are cropped on sandy soils are those within a short distance of the Platte and Arkansas Rivers, which are irrigated. A soil variable not included in maps or analyses is soil depth. The Flint Hills of Kansas, located on the eastern edge of our study area, have fairly fine soil texture; however, these soils are very shallow, and do not support very much cropland. This pattern causes anomalies within the simple relationships among cropland, precipitation, and soil texture (Fig. 8).

Crop Type

In addition to strong correspondence between environmental factors and the presence or absence of cropland, we also observed patterns in the distribution

Distribution of Cropping - Central Great Plains

by Precipitation and Temperature

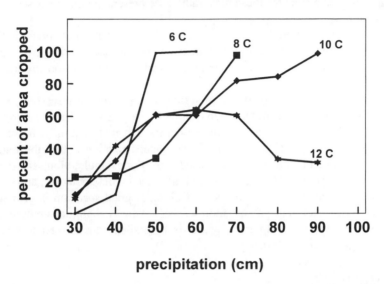

Distribution of Cropping - Central Great Plains

by Precipitation and Sand Content

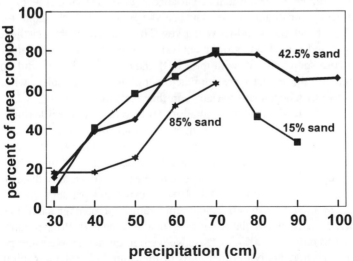

Figure 8. Proportion of area cropped (a) within each mean precipitation–temperature combination and (b) within each precipitation–soil texture combination.

of crop types within the region. Distinctions among crop types on the land use map were made both from spectral characteristics and from knowledge about climatic constraints for crops and crop statistical data. Therefore, we do not present statistical results for these data; rather, we present qualitative assessment of crop patterns.

Areas in western Colorado and Nebraska with less than 500 mm of precipitation support mostly rangeland, but cropped areas within this band fall into two major types. Wheat–fallow agriculture is practiced within this region on productive soils, resulting in alternate years of cropping and soil water storage. In areas that are within approximately 10 km of surface water (rivers or reservoirs) or of aquifers (Nebraska and far southeastern Colorado), irrigated crops may be grown. For our modeling purposes, we ran the model for irrigated corn only; however, there is significant variation in crops grown under irrigation within the region that is not indicated by our remotely sensed map. For example, along the Platte River, corn may be rotated with beans or alfalfa, and onions, potatoes, and sugar beets are grown as well. In southeastern Colorado, grain sorghum is grown in addition to corn. Not until precipitation reaches 70 cm is continuous cultivation as dryland wheat or corn important by area or by grain production (Kansas State Board of Agriculture, 1990).

Statistical Analysis

We conducted a statistical analysis of the importance of physical environmental constraints over land use patterns in the central Great Plains. Our purpose was to obtain an estimate of the amount of variance in land use that can be explained by environmental constraints alone. The analysis was done on the land use data from the USGS Land-Use-Land-Cover survey. These data only classify the region into cropland and rangeland, so our analysis was limited to those categories. These data correspond closely with the EROS data (Fig. 4), but we did not use that map because those data were corrected using climatic variables, which we intended to use as independent variables in this analysis.

We overlaid the land cover, precipitation, temperature, and soil texture maps in the GIS, and created a grid map at a resolution of 1 km. We randomly subsampled 10% of the grid map, providing a data set with 43,234 grid units that could be considered spatially independent of one another. Of this subsample, 60.7% was cropland and 39.3% was rangeland. We then conducted a discriminant analysis [CANDISC procedure (SAS, 1988)], using independent variables of mean annual precipitation, mean annual temperature, percentage sand, and percentage clay, and dependent classes of rangeland or cropland. Prior probabilities were allowed to remain at 50%, because we were interested in assessing general relationships that may apply to other parts of the central Grassland Region, and we did not want to bias the results (Klecka, 1985). All the assumptions required for the discriminant analysis were met with two exceptions. First, interval data were used for the discriminating (independent) variables rather than continuous variables. The discriminating variables from our maps were classed such that there were five temperature classes, ten precipitation classes, and six soil textural classes (for both sand and clay). The midpoints of these classes were used in the analysis. Second, the assumption of no collinearity was not met because of relatively high correlation between sand and clay ($r = -0.66$).

The discriminant analysis indicated that a combination of temperature, precipitation, sand, and clay significantly discriminated between cropland and rangeland areas (Wilks's Lambda test, $p < 0.0001$). Each variable contributed significantly to the discriminant function ($p < 0.0001$). Standardized coefficients were as follows: precipitation, 0.48; temperature, 0.03; sand, -0.73; and clay, -0.02. These coefficients suggest the relative importance of each variable to the function (Klecka, 1985), such that sand and precipitation are far more important in discriminating between rangeland and cropland than are temperature or clay content. The influence of increasing precipitation in the analysis was to increase the discriminant score or likelihood of being classified as cropland; the influence of sand was to decrease the discriminant score and increase the likelihood of classification as rangeland. *A posteriori* classification of the sample resulted in correct classification of 65.7% of the cropland grid units and 68.9% of the rangeland units.

There are several significant factors missing from our analysis with respect to environmental constraints. First, we did not include soil depth as a independent variable; thus, the analysis was unable to explain the presence of rangeland in the Flint Hills of Kansas. Second, we did not try to use a variable such as distance to water supply to assess irrigation patterns. We did not use such a variable because, first, we did not have an aquifer map, and, second, water availability is a complex issue that involves more than just the proximity to a water source. Water availability is determined by availability and expense of technology, and increasingly by water rights and competition between urban users and the agricultural community. Thus, control over patterns of irrigation is not necessarily the result of a physical environmental constraint.

This initial analysis assumed a steady-state distribution of land use. Thus, our assumption for this analysis was that average annual climate variables provide a good index of climatic constraints over land use. In fact, the frequency of marginal years within these average climatic zones may constrain the long-term economic viability of crop production. However, proportional variance in climate across the region is relatively uniform (Lauenroth and Burke, unpublished), such that long-term averages may explain the spatial patterns sufficiently.

The relationship between land use and long-term environmental factors is only representative for a snapshot of land use (actually generated during the early 1980s). Detailed assessments of land use changes in this region suggest that land use has been very dynamic. Recent data for Weld County, CO, in the northeastern corner of the state, indicated several large changes in land use over the last sixty years (Fig. 9). Significant cropland abandonment and government procedure of

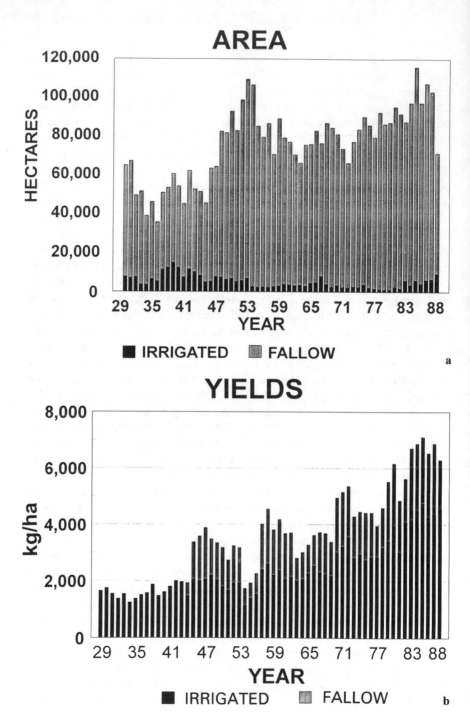

Figure 9. Historical data from Weld County, CO, on (a) acres in wheat production, (b) wheat yield, and (c) total kilograms produced.

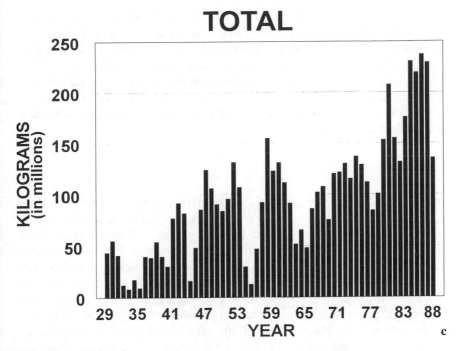

Figure 9. Continued.

wheat–fallow lands occurred during the dustbowl years in the mid 1930s, which were due to very poor precipitation, high winds, and significant crop failure. A long-term increase in acreage under wheat–fallow agriculture occurred between 1929 and 1950, possibly as a result of both increased population in the area and a number of good precipitation years between 1940 and 1950. A dip in acreage under wheat–fallow occurred between 1950 and 1980, probably due to decreased wheat prices, with government programs improving economic viability in the early 1980s. Although we do not have the summary data available, initial county statistics suggest that the Conservation Reserve Program has resulted in significant reductions in wheat–fallow land as the government provides economic incentive for farmers to convert cultivated land back to grasslands. In addition, long-term improvement in crop management systems—including fertilization, reduced tillage, and improved crop strains—create higher grain yields and reduce impacts on soil organic matter (Fig. 9b). Regional-scale patterns in acreage under cultivation and technological improvements have thus resulted in large-scale dynamic effects on regional grain yield (Fig. 9c), as well as potential effects of land use management practices on ecosystem carbon storage.

Long-term patterns in physical environmental constraints are limited in their power to explain land use patterns in the central Great Plains of the United States. Our discriminant function does not have the ability to predict future changes in land use. First, our assessment suggests that environmental constraints only explain about two-thirds of the variance in spatial patterns of land use. The important missing factors in this analysis are socioeconomic. Factors such as family tradition, local economic conditions and quality of life, regional/continental and global markets, government support programs, and technological advances must be considered to assess trends in land use changes adequately (Riebsame et al., 1994). Such predictions are critical for long-term assessment of human–environmental interactions.

Summary

Human activities interacting with physical environmental constraints have and will continue to have important effects on the structure and function of ecosystems in the Central Grassland region of the United States. Conversions of native grasslands to croplands have had enormous effects on the prosperity of the region but not without ecological costs. Cultivation has resulted in absolute losses of soil organic matter and nutrients. Simulations to evaluate the magnitudes and geographic distribution of such losses in the central Great Plains suggested that on average, across all kinds and locations of cultivated fields, 44% of the total soil organic matter was lost in the first fifty years of cultivation. Geographically, the largest losses occurred in areas with the highest original stores of soil organic matter. These occurred in the areas with fine texture soils and the highest amounts of annual precipitation.

There is an important ecological relationship between the environmental factors that result in large losses of soil organic matter following cultivation and those factors that determine whether a particular location will be cultivated or not. Statistical analysis of the relationship between the presence of cropland and environmental factors indicated a positive relationship with precipitation amount and a negative relationship with sand content of the soil. A qualitative analysis of our simulation results indicates that soil organic matter losses are also positively related to precipitation amount and negatively related to the sand content of soil.

The interrelationships between the geographic distributions of cropland and soil organic matter losses are testimony to the importance of physical environmental controls in influencing land use and its ecological consequences. However, the fact that our statistical analysis left the location of one-third of the cropland unexplained emphasizes the importance of social, economic, and cultural factors as a modifier of simple environmental decisionmaking.

Acknowledgments

The authors wish to thank Martha Coleman for geographic analysis and graphic production. We also thank Tammy Bearly for programming and simulation analysis, and Alister Metherill for providing parameter estimates for the CENTURY model. P. Hook and M. A. Vinton provided reviews of the manuscript. Support for this work was provided by grants from the National Science Foundation (BSR # 91-06183 and BSR 90-11659), the USDA Agricultural Research Service, and the Colorado State Agricultural Experiment Station (1-50661).

References

Aguilar, R., E. F. Kelly, and R. D. Heil. 1988. Effects of cultivation on soils in northern Great Plains rangelands. *Soil Science Society of America Journal* **52:**1076–1081.

Anderson, J. R., E. E. Hardy, J. T. Roach, and R. E. Witmer. 1976. A land use and land cover classification system for use with remote sensor data. *U.S. Geological Survey Professional Paper 964.*

Borchert, J. R. 1950. The climate of the central North American grasslands. *Annals of the Association of American Geographers* **40:**1–39.

Bormann, F. H., and G. E. Likens. 1979. *Pattern and Process in a Forested Ecosystem.* Springer-Verlag, New York.

Burke, I. C., C. M. Yonker, W. J. Parton, C. V. Cole, K. Flach, and D. S. Schimel. 1989. Texture, climate, and cultivation effects on soil organic matter content in U.S. grassland soils. *Soil Science Society of America Journal* **53:**800–805.

Burke, I. C., T. G. F. Kittel, W. K. Lauenroth, P. Snook, C. M. Yonker, and W. J. Parton. 1991. Regional analysis of the central Great Plains. *BioScience* **41:**685–692.

Burton, I., R. W. Kates, and G. F. White. 1978. *The Environment as Hazard.* Oxford University Press, New York.

CLIMATEDATA. 1988. U.S. West Optical Publishing, Denver, CO.

Doran, J. W. 1980. Soil microbial and biochemical changes associated with reduced tillage. *Soil Science Society of America Journal* **44:**518–24.

Dregne, H. E., and W. O. Willis. 1983. *Dryland Agriculture.* American Society of Agronomy, Madison, WI.

Elliott, E. T., and C. V. Cole. 1989. A perspective on agroecosystem science. *Ecology* **70:**1597–1602.

Emanuel, W. R., H. H. Shugart, and M. P. Stevenson. 1985. Climate change and the broad-scale distribution of terrestrial ecosystem complexes. *Climate Change* **7:**29–43.

Haas, H. J., C. E. Evans, and E. R. Miles. 1957. Nitrogen and carbon changes in soils as influenced by cropping and soil treatments. *USDA Technical Bulletin* 1164. U.S. Government Printing Office, Washington, DC.

Hide, J. C., and W. H. Metzger. 1939. The effect of cultivation and erosion on the nitrogen and carbon of some Kansas soils. *Agronomy Journal* **31:**625–632.

Holland, E. A., and D. C. Coleman. 1987. Litter placement effects on microbial and organic matter dynamics in an agroecosystem. *Ecology* **68:**425–433.

Holdridge, L. R. 1947. Determination of world plant formations from simple climatic data. *Science* **105**:367–368.

Houghton, R. A. 1990. The future role of tropical forests in affecting the carbon dioxide concentration of the atmosphere. *Ambio* **19**:204–209.

Houghton, R. A., R. D. Boone, J. M. Melillo, C. A. Palm, G. M. Woodwell, N. Myers, B. Moore, and D. L. Skole. 1987. Net flux of CO_2 from tropical forests in 1980. *Nature* **316**:617–620.

Jenkinson, D. S. 1977. Studies on the decomposition of plant material in soil. V. The effects of plant cover and soil type on the loss of carbon from 14-C labelled rye grass decomposing under field conditions. *Journal of Soil Science* **28**:424–494.

Jenny, H. 1930. A study on the influence of climate upon the nitrogen and organic matter content of the soil. *Montana Experiment Station Bulletin 152*.

Jenny, H. 1980. *The Soil Resource*. Springer-Verlag, New York.

Kansas State Board of Agriculture. 1990. *Kansas Farm Facts*. Kansas Agricultural Statistics and USDA National Agricultural Statistics Service, Topeka, KS.

Klecka, N. 1985. *Discriminant Analysis*. Sage Publications, Newbury Park, N.J.

Loveland, T. R., J. W. Merchant, D. O., Ohlen, and J. F. Brown. 1991. Development of a land-cover characteristics database for the conterminous U.S. *Photogrammetric Engineering and Remote Sensing* **57**:1453–1463.

Lowenthal, D. 1961. Geography, experience, and imagination: towards a geographical epistemology. *Annals of the Association of American Geographers* **51**:241–260.

Lowenthal, D. 1972. The nature of perceived and imagined environments. *Environment and Behavior* **4**:189–207.

McArthur, R. H. 1972. *Geographical Ecology*. Harper and Row, New York.

Metherill, A. 1992. Application of an ecosystem simulation model to cropping systems in the Central Great Plains. Dissertation, Colorado State University.

Meyer, W. B., and B. L. Turner II. 1992. Human population growth and global land-use/cover change. *Annual Reviews* **23**:39–61.

Muhs, D. R. 1985. Age and paleoclimatic significance of Holocene sand dunes in northeastern Colorado. *Annuals of the Association of American Geographers* **75**:566–582.

Parton, W. J., D. J. Schimel, C. V. Cole, and D. S. Ojima. 1987. Analysis of factors controlling soil organic matter levels on grasslands. *Soil Science Society of America Journal* **51**:1173–1179.

Paul, E. A. 1984. Dynamics of organic matter in soils. *Plant and Soil* **76**:275–285.

Pielke, R. A., and R. Avissar. 1990. Influence of landscape structure on local and regional climate. *Landscape Ecology* **4**:133–155.

Post, W. M., T. H. Peng, W. R. Emanuel, A. W. King, V. H. Dale, and D. L. DeAngelis. 1990. The global carbon cycle. *American Scientist* **78**:310–326.

Riebsame, W. E., K. A. Galvin, R. Young, W. J. Parton, I. C. Burke, L. Bohren, and E. Knop. 1994. An integrated model of causes and responses to environmental change. *BioScience*. In press.

Sala, O. E., W. J. Parton, L. Joyce, and W. K. Lauenroth. 1988. Primary production of the central grassland region of the United States. *Ecology* **69**:40–45.

SAS Institute, Inc. 1988. *SAS/STAT User's Guide, Release 6.03 ed.* SAS Institute, Inc. Cary, NC.

Schlesinger, W. H. 1990. Evidence from chronosequence studies for a low carbon-storage potential of soils. *Nature* **348**:232–234.

Sorenson, L. H. 1981. Carbon-nitrogen relationships during the humification of cellulose in soils containing different amounts of clay. *Soil Biology and Biochemistry* **13**:313–321.

Stern, P. C., O. R. Young, and D. Druckman (eds.). 1992. *Global Environmental Change. Understanding the Human Dimensions.* National Academy Press, Washington, D.C.

Tiessen, H., J. W. B. Stewart, and J. R. Bettany. 1982. Cultivation effects on the amounts and concentrations of carbon, nitrogen, and phosphorus in grassland soils. *Agronomy Journal* **74**:831–875.

USDI U.S. Geological Survey. 1986. Land use and land cover digital data from 1:250,000 and 1:100,000 scale maps. *National Mapping Program Technical Instructions, Data Users Guide* 4, Reston, VA.

USDA Soil Conservation Service. 1989. *STATSGO Soil Maps.* National Cartographic Center, Fort Worth, TX.

Watt, A. S. 1947. Pattern and process in the plant community. *J. Ecology* **35**:1–22.

Whittaker, R. H. 1953. A consideration of climax theory: the climax as a population and pattern. *Ecological Monographs* **23**:41–78.

Whittaker, R. H. 1973. Climax concepts and recognition. *Handbook of Vegetation Science* **8**:137–154.

7

The Study of Ozone Climatology and Pollution in the Northeastern and Southern United States Using Regional Air Quality Models

William L. Chameides

Introduction

Photochemical smog and its attendant high concentrations of ozone pollution were first identified as an environmental problem in Los Angeles in the 1950s (Haagen-Smith, 1952). (Ironically, while ozone in the stratosphere is believed to protect living organisms from harmful ultraviolet radiation, ozone at the earth's surface is generally thought of as a pollutant because this strongly oxidizing gas can damage living tissue by direct contact.) Today, in spite of extensive research and, in some cases, large expenditures of funds for pollution abatement, photochemical smog is not only a serious problem in Los Angeles, but in virtually every major urban center in the world. In addition to being an urban problem, there is growing evidence that photochemical smog poses a threat to ecosystems in many rural areas. This threat appears to be especially chronic in the northeastern and southern United States, where a warm, stagnant summertime climatology combines with ample emissions of natural hydrocarbons and anthropogenic nitrogen oxides to produce high concentrations of ozone throughout the region during the summer months. In this chapter, a review is presented of the ozone pollution problem in the northeastern and southern United States and the use of regional air quality models to simulate the development of this pollution and ultimately develop control strategies for its abatement.

What is Photochemical Smog and Ozone Pollution?

Photochemical smog refers to the mix of gases and particles produced in the Earth's lower atmosphere or troposphere by chemical reactions involving either VOC (i.e., hydrocarbons and other volatile organic compounds) and NOx (NO + NO_2) in the presence of sunlight (see Fig. 1). In addition to VOC, carbon monoxide (CO) can also react with NOx in the presence of sunlight to produce photochemical smog. Of particular concern in photochemical smog are the high

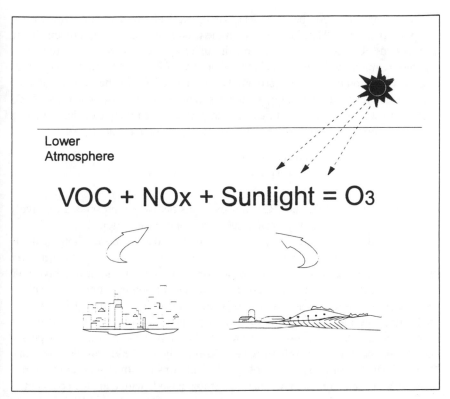

Figure 1.   Ozone in the lower atmosphere or troposphere is produced from photochemical reactions involving VOCs (volatile organic compounds) and NOx (nitrogen oxides) in the presence of sunlight.

concentrations of ozone (O_3) that typically occur. While uncertainties remain in our understanding of tropospheric photochemistry, the basic set of reactions that lead to O_3 production have been identified. These reactions, commonly referred to in the aggregate as the "photochemical smog mechanism," involve the oxidation of hydrocarbons and other volatile organic compounds in the presence of nitrogen oxides (NOx) and sunlight (Haagen-Smit, 1952; Seinfeld, 1989). Typical of this mechanism are Reactions (R1) through R7):

$$
\begin{aligned}
\text{(R1) } RH + OH &\rightarrow R + H_2O \\
\text{(R2) } R + O_2 + M &\rightarrow RO_2 + M \\
\text{(R3) } RO_2 + NO &\rightarrow RO + NO_2 \\
\text{(R4) } RO + O_2 &\rightarrow HO_2 + R'CHO \\
\text{(R5) } HO_2 + NO &\rightarrow OH + NO_2 \\
2X \text{ (R6) } NO_2 + h\upsilon &\rightarrow NO + O \\
2X \text{ (R7) } O + O_2 + M &\rightarrow O_3 + M
\end{aligned}
$$

$$
\text{NET: } RH + 4O_2 + 2h\upsilon \rightarrow R'CHO + H_2O + 2O_3
$$

where an initial reaction between RH (used generically to denote a hydrocarbon compound) and an OH radical results in the production of two O_3 molecules and an aldehyde R'CHO. (The prime on the aldehyde is used here to indicate an organic fragment with one less C atom than that of R.) Additional O_3 molecules then can be produced from the degradation of R'CHO. In the case of olefinic hydrocarbons, ozone-producing sequences can be initiated by reactions of RH with O_3 as well as with OH. Finally, it should be noted that, like the oxidation of hydrocarbons, O_3 can be generated from CO oxidation via

$$(R8)\ CO + OH \rightarrow CO_2 + H$$
$$(R9)\ H + O_2 + M \rightarrow HO_2 + M$$

followed by (R5), (R6), and (R7); however this pathway represents a relatively minor source of O_3 in many urban, suburban, and rural regions.

While VOC and NOx are required to produce ozone from the photochemical smog mechanism, high concentrations of VOC and NOx do not by themselves guarantee the generation of high ozone concentrations. In addition, favorable meteorological conditions are needed. These meteorological conditions include clear skies, warm temperatures and stagnant wind conditions (most often associated with slow-moving, summertime high pressure systems) that speed up the photochemical smog reactions and prevent the pollutants from being dispersed into the background atmosphere (Chen, 1989; Logan, 1989; van den Dool and Saha, 1990). When these meteorological conditions combine with ample emissions of VOC and NOx, a photochemical smog episode with enhanced concentrations of ozone near the Earth's surface often ensues. Depending on the movement of the high-pressure system, such episodes can persist for as little as a day or two, to as long as a week or more.

Because ozone is a strong oxidant, the high concentrations of this compound typically encountered during photochemical smog episodes can be harmful to humans and other living organisms (see Table 1). In addition to ozone, other oxidants such as peroxides, as well as acidic aerosols and eye irritants such as peroxyacetyl nitrate (PAN), are also associated with photochemical smog.

In urban areas with large sources of VOC and NOx from pollution, photochemi-

Table 1. *Effects of Ozone From Photochemical Smog*

Human Health
- Reduction in lung function (brief exposure)
- Possible permanent lung damage (chronic exposure)

Welfare
- Reduced crop yield
- Reduced tree growth
- Possible enhancement in global Greenhouse Effect

cal smog reactions driven by intense summer sun and high temperatures often produce summertime ozone concentrations that can be harmful to human health. In the United States, the Environmental Protection Agency (EPA), under mandate of the U.S. Congress, first established a National Ambient Air Quality Standard (NAAQS) for all atmospheric oxidants of 0.08 ppmv in 1971; this was later revised to the current NAAQS for ozone alone of 0.12 ppmv in 1979. (Note 0.12 ppmv = 0.12 parts per million by volume or approximately one molecule of ozone for every 8 million molecules of air.) On the basis of numerous scientific studies, it is now well established that a significant fraction of a population exposed to ozone in excess of the NAAQS will experience acute reductions in lung functions [see Lippmann (1989)]. People affected by ozone pollution include normal healthy adults and children as well as those with impaired respiratory systems. While there is some evidence to suggest that repeated exposures to high ozone concentrations cause permanent lung damage (for example, emphysema or lung cancer), this evidence is not conclusive and thus the long-term consequences of ozone exposures are not yet certain.

While exceedances of the NAAQS for ozone occurs most commonly in urban areas, problems associated with photochemical smog and high concentrations of ozone are by no means limited to the urban environment. High concentrations of ozone, sometimes in excess of the NAAQS, have also been observed in rural areas, especially in the United States and Europe (Derwent, 1989; Hov et al., 1978; Isaksen et al., 1978; LeFohn and Pinkerton, 1988; Logan, 1989). Although this photochemical smog problem has a much smaller impact on human health because of the lower ozone concentrations that typically occur in rural areas and the lower population densities of rural areas, the ecological and economic impact of rural smog can be quite significant. Scientific studies have demonstrated conclusively that repeated exposure to ozone causes a depression in the growth rate and ultimate yield from a wide variety of agricultural crops and trees (Adams et al., 1989; Heck et al., 1982). Ozone-sensitive species of economic importance include soybeans, cotton, peanuts, clover, beans, potatoes, watermelon, tomatoes, squash, and radishes as well as southern pines and oak. In general, ozone concentrations do not have to be in excess of NAAQS to cause damage to vegetation, and deleterious effects have been documented for ozone concentrations as low as 0.05 ppmv (Heck et al., 1982). Because rural ozone concentrations of this magnitude are not uncommon in the United States (see Fig. 2), the impact is likely to be significant. In fact, the recently completed study by the National Crop Loss Assessment Network (Adams et al., 1989) concluded that ozone pollution in rural areas of the United States is currently causing an approximate 10% loss in agricultural productivity at a cost of about $5 billion per year and may also be a significant factor in forest declines occurring over broad areas of the nation.

The effects of photochemical smog may not be limited to those associated with declining air quality. Rising concentrations of ozone produced from photo-

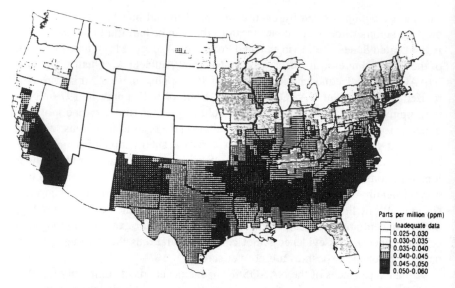

Figure 2. Estimated daily 7-hour average ozone concentration during the growing season over the United States [from EPA (1991)].

chemical smog in urban and rural areas of the United States appear to be part of a larger pattern of ozone increases that may be occurring throughout much of the Northern Hemisphere (Volz and Kley, 1988). Since tropospheric ozone is a greenhouse gas, these rising concentrations may be a contributor to a global warming caused by an enhanced Greenhouse Effect (IPCC, 1990).

Photochemical Smog in the Northeastern and Southern United States

Like most regions of the nation, the lands of the northeastern and southern United States are dedicated to a variety of human activities and uses. In general, the northeastern United States tends to be more industrialized and urbanized, while a major fraction of the land area in the South is rural and is currently used for agriculture or is covered by forests. However, in spite of the different characteristics of the regions, both regions, at least superficially, appear to have a similar problem with photochemical smog pollution on the regional scale. As illustrated in Figures 2 and 3, the northeastern and southern United States are subjected to enhanced concentrations of ozone over the summer months, with large portions of both regions exposed to daytime ozone concentrations in excess of 50 ppbv over much of the growing season. The highest ozone concentrations appear to occur during specific multiday regional episodes with high ozone concentrations and haze over much of the northeast and south (see Fig. 4). These episodes are characterized by slow-moving, high-pressure systems that bring ample sunshine,

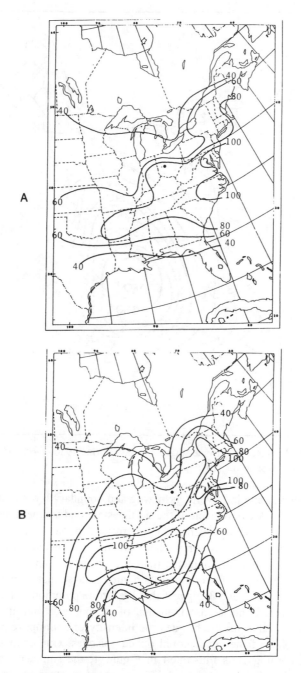

Figure 3. Synoptic-scale analysis of the diurnal maximum ozone concentration (ppbv) for (A) August 2, 1980 and (B) August 8, 1980. [From Vukovich et al. (1985)].

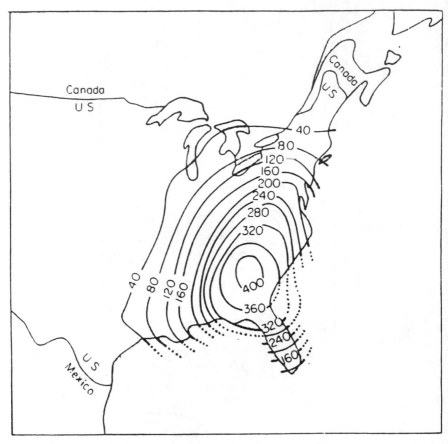

Figure 4. Total number of extreme stagnation days during 1936–1965 east of the Rocky Mountains. [After Korshover (1976)].

high temperatures, and little vertical mixing and boundary layer venting (Vuko-vich et al., 1985). A comparison of the northeastern and southern regions of the United States raises an interesting dichotomy and points to the complexities inherent in the phenomenon of photochemical smog. As noted earlier, the south tends to be more rural in character than the northeast and yet both regions experience similar levels of ozone pollution during the summer months. (In fact, inspection of Figures 2 and 3 suggest that the ozone pollution in the south may be more widespread and intense than that of the northeast and the rest of the nation.) Clearly, other factors besides the intensity of anthropogenic emissions must play a role in fostering regional ozone episodes. One of these factors is certainly meteorology. As illustrated in Figure 4, the summer months in the south tend to be characterized by highly stagnant weather patterns with low winds, high temperatures, and ample sunshine—the conditions most conducive to the accumulation of air pollutants and the generation of photochemical smog. While stagnation also occurs in the northeast, it appears to be less common than in the south. Support for this conclusion can be found in the work of Samson and Shi (1988), who found that high ozone episodes in northeastern cities appear to be associated more with pollutant transport from distant sources than similar episodes in southern cities.

Another factor affecting photochemical smog may be natural hydrocarbon emissions. As noted below, the southern United States has relatively high rates of natural hydrocarbon emissions and these emissions tend to be dominated by emissions of isoprene, a highly reactive hydrocarbon. Unfortunately for the southern United States, it would appear that the same characteristics that make the region favorable for agriculture and silviculture also make it an environment that is conducive to photochemical smog production.

Understanding the causes of the high ozone concentrations in the northeastern and southern United States and developing a regional strategy for the abatement of these ozone concentrations are confounded by two additional elements. The first of these is the complex nature of the photochemical smog mechanism itself. As noted earlier, virtually all ozone in the lower atmosphere is the product of chemical reactions involving VOC and NOx. (A contribution is also made from reactions involving CO and NOx.) The NOx is emitted from high-temperature combustion processes, primarily in the form of NO. The VOC originate from a variety of natural and human-source emissions and includes a wide range of different compounds, each having a characteristic efficiency of producing ozone when exposed to sunlight.

In principle, therefore, one might conclude that the control of ozone pollution could be accomplished by lowering emissions of VOC, or NOx, or both. However, the production of ozone from photochemical smog has a very nonlinear dependence on the concentrations of NOx and the various VOC present in the atmosphere. This nonlinear dependence is illustrated in Figure 5, where model-calculated isopleths of peak O_3 concentrations are plotted as a function of assumed total VOC and NOx concentration. Note that the effect of O_3 of reducing VOC and NOx changes considerably as the concentrations of VOC and NOx change. When VOC levels are relatively high and NOx levels are relatively low (i.e., in the lower right-hand portion of the figure), O_3 production is limited by the availability of NOx and reductions in VOC have little or no effect on O_3. Conversely when hydrocarbon levels are low and NOx levels are high (i.e., the upper left-hand corner of the diagram), O_3 production is limited by hydrocarbons and NOx reductions are ineffective. Thus, in order to devise an effective strategy for O_3 abatement in a given airshed, it is first necessary to determine where on the ozone-isopleth diagram that airshed is located. In the northeastern United States, determination of whether a VOC-based or NOx-based strategy is further complicated by the likelihood that some portions of the region may be VOC-limited while others may NOx-limited (NAS, 1991).

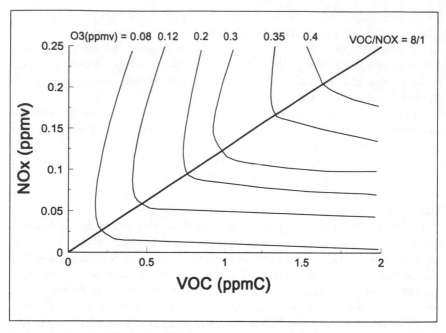

Figure 5. Model-calculated peak ozone concentrations as a function of assumed VOC and NOx concentrations.

The second confounding element in devising an ozone-abatement strategy arises from the presence of natural VOC in the atmosphere of the northeastern and southern United States. Emission studies as well as actual observations of atmospheric VOC concentrations in the atmosphere indicate that vegetation in general and trees in particular emit large quantities of reactive hydrocarbons during the growing season (Chameides et al., 1992; Lamb et al., 1987; Zimmerman, 1979). In fact, natural hydrocarbon emissions appear to be the dominant source of VOC to the atmosphere in rural areas of the northeastern and southern United States and a significant source in many urban areas. (Natural hydrocarbon emissions appear to be especially large in the southern United States where warm summertime temperatures and extensive forests favor biogenic hydrocarbon emissions.) These large natural emissions make ozone abatement even more difficult because these emissions cannot be controlled. As a result, hydrocarbon emission controls may not be effective in the northeastern and southern United States, even at times when the atmosphere is in the hydrocarbon-limited region of the ozone-isopleth diagram (Chameides et al., 1988; Trainer et al., 1987). The development of an effective strategy is thus also dependent on a reliable assessment of the relative contributions of anthropogenic and natural precursors to the airshed.

Most likely because of these complexities, the implementation of urban ozone-abatement strategies in the United States has had very limited success. In the late 1970s it was determined that a VOC-based strategy would be most effective, and regulations were promulgated to limit VOC emissions from mobile and stationary sources. Today, some fifteen years later, little progress in reducing urban ozone in the United States has occurred, and there is growing evidence that a NOx-based strategy will be needed to bring the nation's cites into compliance with EPA's NAAQS (NAS, 1991).

Because of the complexities described above, an analysis of the causes of photochemical smog formation in the northeastern and southern United States and its control ultimately requires the use of mathematical models capable of simultaneously accounting for the myriad of processes and factors that control the photochemical smog system. One specific form of mathematical model, the air-quality model, attempts to simulate the relevant physical and chemical processes using mathematical equations. Because of the complex and nonlinear nature of these equations, air-quality models usually use numerical techniques for solving these equations, hence, the term "computer models." The application of an air-quality model to a regional problem such as photochemical smog in the northeastern and southern United States often is referred to as the application of a "regional-air-quality model." In the next section the basic features of regional-air-quality models and their application to photochemical smog in the northeastern and southern United States are reviewed.

The Regional Air-Quality Model—Main Attributes

The fundamental challenge of all air-quality models is to describe in quantitative terms the so-called "source–receptor relationships" that tie sources or emissions of gaseous or particulate species at one location to the subsequent appearance and removal of these same or related species at some other location (see Fig. 6). In the case of photochemical smog, this challenge is made more complicated by the fact that the species of greatest concern, such as ozone, are secondary pollutants; that is, pollutants produced from chemical reactions of the primary pollutants VOC and NOx. Thus, an air-quality model capable of describing the relationship between VOC and NOx sources and the effects of photochemical smog and its related ozone pollution must accurately simulate the relevant photochemical tranformations that take place as well as the transport by winds and turbulence of the primary and secondary pollutants.

Air-quality models include simple box models that neglect transport and only consider photochemical processes and lagrangian or trajectory models that follow a specific parcel of air as it moves over the Earth's surface (Seinfeld, 1989; Tesche, 1983). However, the most sophisticated of the air-quality models generally are considered to be the eulerian or grided models. In these models the three-dimensional modeling domain is divided into a grid of homogeneous boxes

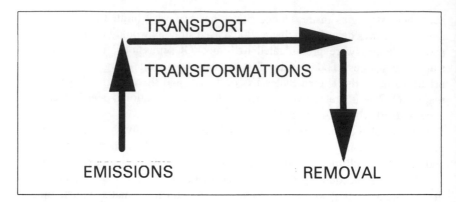

Figure 6. The challenge of air-quality models is to determine source/receptor relationships; that is, the connection between pollutant emissions, their transport and photochemical transformation into secondary pollutants, and their eventual removal via deposition.

of a given dimension (see Fig. 7) and the concentrations of chemical species within each grid are determined by the model as a function of time. Examples of three-dimensional regional, eluerian models include the Regional Acid Deposition Model or RADM (Chang et al., 1987; Middleton et al., 1988), the Acid Deposition and Oxidant Model or ADOM (Venketram et al., 1988), EPA's Regional Oxidant Model or ROM (Schere and Wayland, 1989), NOAA's regional oxidant model (McKeen et al., 1991), and STEM (Charmichael and Peters, 1984). While there are differences in the detailed structure, mechanisms, and numerics of each of these models, they have in common two important features: (1) They are currently used in an episodic mode; that is, they attempt to simulate a specific pollution episode rather than the climatology of photochemical smog in a region. (2) They are all designed to solve a system of continuity equations using numerical techniques, as described below.

Eulerian air-quality models determine source–receptor relationships by solving a series of coupled differential equations that describe the time rate of change of the relevant species' concentrations at each point in space (that is, within each grid). These equations, called continuity equations, are mathematical statements of conservation of mass and have the following general form:

Time rate of change of species concentration within volume centered at point (xyz) = {Production of species in volume} − {Loss of species in volume} + {Transport of species into volume} − {Transport of species out of volume}

In general, the production and loss terms represent the contributions from photochemical processes; however, for volumes located at the surface, these terms are also used to represent emissions of the species from surface sources and losses of the species due to deposition on the surface, respectively.

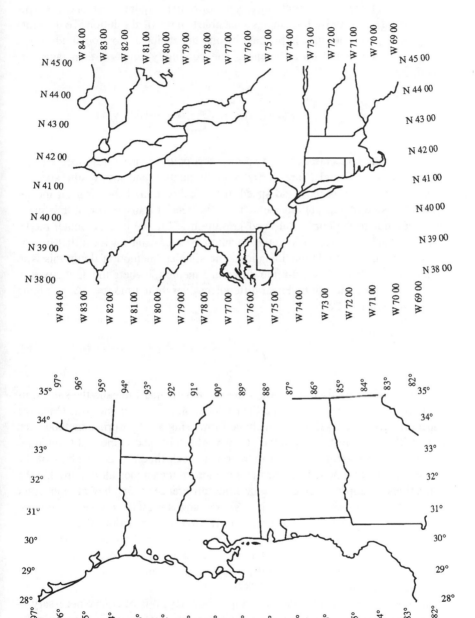

Figure 7. Model domain for ROM application to (A) the northeastern United States and (B) the southeastern United States [After Schere and Wayland (1989)].

Solution of the above continuity equations first requires that one cast this equation in a quantitative form that is amenable to numerical solution. To illustrate this process, consider the simple case of one-dimensional, horizontal advection in the x direction and no turbulence. The continuity equation then takes the following form:

$$\frac{\partial C(x,y,z,t)}{\partial t} = P(x,y,z,t) - L(x,y,z,t) - \frac{\partial}{\partial x}vC(x,y,z,t) \tag{1}$$

where C is the concentration of a species at a point in space described by x, y, and z at time t; P and L are the rates of production and loss of the species, respectively; and v is the wind speed in the x direction. [The term on the far right-hand side of Eq. (1) represents the net effect of transport or, in this case, advection due to the horizontal wind.] Solution of Eq. (1) by a computer model requires that the above differential equation be recast into a form that can be solved through algebraic manipulation. The simplest method of doing this is to adopt the so-called "forward-finite difference method," where the derivatives in Eq. (1) are approximated by linear differences over finite time steps Δt and finite spatial increments Δx:

$$\frac{C_{x,y,z}^{t+\Delta t} - C^t}{\Delta t} = (P_{x,y,z}^t - L_{x,y,z}^t) - \frac{(v_{x+\Delta x,y,z}^t C_{x+\Delta x,y,z}^t - v_{x-\Delta x,y,z}^t C_{\Delta x,y,z}^t)}{2\Delta x} \tag{2}$$

The superscripts are used to denote values at a particular time and the subscripts are used to denote values at a particular location on the spatial grid. Once the equations have been written in algebraic form using an appropriate method such as that illustrated in Eq. (2), a variety of well-established numerical techniques can be used to solve for the concentrations at each time $t + \Delta t$ in terms of the concentrations at time t. Because these numerical techniques all require that the derivatives be approximated in some finite-difference form, they all encounter some degree of truncation error. These errors can manifest themselves as nonconservation of mass (that is, the artificial creation or destruction of the species) and/or numerical diffusion (that is, the smearing out of the species distribution in space). To minimize these problems, investigators have developed a variety of higher-order schemes that limit truncation errors and filtering schemes that limit numerical diffusion (Rood, 1987).

In order to use the above-described methodology, it is of course necessary to first determine and/or specify the various parameters that appear in the continuity equations within each grid of the model. These parameters include emissions from the surface, winds and related meteorological parameters, photochemical transformation rates, and deposition rates. Each of these parameters is discussed briefly below.

Emissions

The surface sources specified in air-quality models are based typically on so-called "emission inventories." In these inventories, the rate of emission of a species from a given source category (mobile sources, power plants, biogenic sources, etc.) is estimated from two factors: (1) an empirically-derived Emission Factor, EF, which relates the amount of emission from a single unit source (that is, a single car for automotive sources, a single power plant for point sources, or a unit mass of a tree species in the case of biogenic emissions) and (2) an activity level Factor, ALF, which specifies the number of unit emitters present within any grid or spatial domain. Thus,

$$\text{Emission Rate} = \text{ALF} \bullet \text{EF} \qquad (3)$$

In general, the specification of the Emission Factor can lead to large uncertainties in the estimated Emission Rate because of the variability that can occur between like sources. For instance, two identical model cars can have vastly different pollutant emission rates because of differences in the maintenance of the two engines (NAS, 1991). This becomes an especially difficult problem when attempting to run the model in an episodic mode because, in this case, the emission factors can depend on day-to-day variability in energy demands, traffic patterns, and photosynthetic rates.

The specification of the Activity Level, on the other hand, represents a relatively straightforward procedure of inventorying the various sources (hence, the expression emission inventory). For some sources, however, particularly those that are relatively small and distributed throughout an area, this task can become quite arduous and fraught with potential errors.

For these reasons large uncertainties (of the order of a factor of 2 or more) typically are associated with the emissions input into air-quality models, and these uncertainties of course propagate throughout the simulation and can, in general, lead to significant errors in the model results. This is a particularly worrisome problem in the case of photochemical smog simulations because of the nonlinearities inherent in the formation mechanism. For this reason, results from air-quality models should in general be compared carefully with more independent observation-based analyses that do not depend on emission inventories (Chameides et al., 1992).

Winds and Meteorological Parameters

Specification of three-dimensional wind fields can be derived from three basic sources: (1) observed fields; (2) predicted fields from prognostic meteorological model calculations (Anthes and Warner, 1978); and (3) a combination of sources in which observed fields are combined with prognostic fields in a so-called four-dimensional data assimilation mode or as input into a diagnostic model that determines a hybrid field based on mass conservation. All approaches have advantages and disadvantages. The use of observed fields is in many respects the simplest and least taxing on computational resources but requires a detailed observational network and can introduce significant inaccuracies in areas with complex terrain. Prognostic model calculations can in principle simulate all relevant meteorological features at the scale required based on a specification of the large-scale flow, terrain features, and boundary conditions, and, in that sense are advantageous relative to using observed fields. On the other hand, prognostic models are computer intensive and thus can tax available computer resources; more important, prognostic models cannot be expected to reproduce exactly observed wind and related meteorological fields.

Photochemical Processes

The accurate simulation of photochemical processes is clearly a key component of any air-quality model attempting to simulate photochemical smog formation. However, the relevant photochemical processes are extremely complex, and, as a result, all models must adopt approximate rather than exact mechanisms. The atmosphere contains hundreds of different anthropogenically derived and biogenically derived VOCs. Each VOC can in principle have a unique chemical pathway for its oxidation. Because of uncertainties in these individual chemical pathways and limitations on computer resources, it is unrealistic to attempt to simulate the chemistry of each and every VOC in the atmosphere. Instead air-quality models typically utilize semiempirical mechanisms in which a relatively small number (about ten) model-VOC compounds are used to simulate the full suite of VOCs actually found in the atmosphere. These mechanisms use either a "lumped" approach in which like VOCs are lumped together or a "surrogate" approach in which a specific VOC is used as a surrogate for a whole class of VOCs. Once a basic approach has been established, the mechanism then is "tuned" to reproduce characteristic photochemical smog formation rates as observed in laboratory experiments using smog chambers. Examples of these chemical mechanisms include the Carbon Bond Mechanism (Gery et al., 1989), the CAL mechanism (Carter et al., 1986; Lurmann et al., 1986), and the RADM mechanism (Stockwell et al., 1990). A detailed comparison by Dodge (1989) of the predicted smog formation rates for each of these mechanisms indicated reasonably good agreement (see Fig. 8). However, significant uncertainties remain in the use of these mechanisms because of the difficulties in interpreting smog chamber data and the possibility that smog chamber data may not, in all cases, be directly applicable to the atmosphere.

Surface Deposition

Deposition of chemical species onto the Earth's surface, especially in the case of oxidizing chemicals such as O_3, can represent a significant loss pathway. In

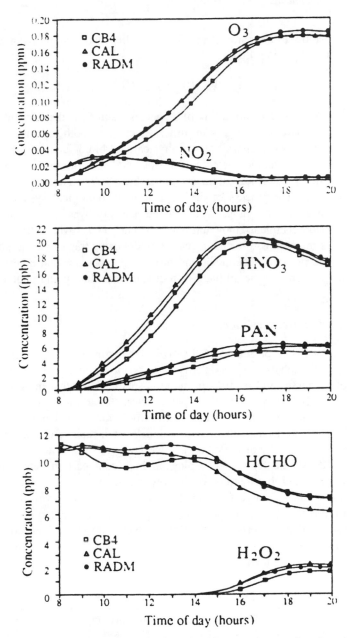

Figure 8. Predicted concentrations of selected oxidant species as a function of time for a specific case scenario designed to simulate conditions in Philadelphia with an initial VOC concentration of 0.54 ppmC and an initial NOx concentration of 0.06 ppmv. CB4 indicates results using the carbon bond mechanism, CAL indicates results using the CAL mechanism, and RADM indicates results using the RADM mechanism (see text). [From Dodge (1989)].

general, these fluxes are parametrized in air-quality models in terms of a deposition velocity, V_{dep}, such that

$$\Phi_{dep}(I) = C_I V_{dep}(I) \qquad (4)$$

where $\Phi_{dep}(I)$ is the deposition flux of a species I and C_I is the concentration of species I. The parameter V_{dep} is in turn a function of both the meteorological state of the atmosphere as well as the physical and biological characteristics of the surface soil, vegetation, etc., and is specified in terms of relatively well-developed micrometeorological theory (Runeckles, 1992). However, it is important to note that the process of dry deposition to biological systems such as vegetation or a human's lungs is in fact the pathway by which the deleterious effects of photochemical smog ultimately occur. Thus, while dry deposition typically is simulated within an air-quality model using this single parameterization, V_{dep}, the physical, chemical, and biological processes that determine V_{dep} are key components of the problem of regional air pollution and photochemical smog.

Subgrid Scale Parameterizations

In Table 2, a list is presented of the typical horizontal grid sizes for urban-, regional-, and global-scale models as well as the typical spatial scales for a variety of processes and features in the atmosphere. Note that the grid size for regional models is larger than the spatial scales for a variety of atmospheric features (including turbulent eddies, convective clouds, and power plant and urban plumes) and, as a result, regional models are incapable of resolving these processes. The inability of regional models to resolve these features is a serious problem because the features can play an important part in determining the formation rates and distributions of the key species involved in photochemical smog. For instance, convective clouds can be a major pathway for injecting pollutants into the free troposphere [see Fig. 9 from Dickerson et al. (1987)] and failure to treat properly the dispersal of NOx emissions from power plant and urban plumes can lead to a significant overestimate in ozone formation rates [see Fig. 10 from Sillman et al. (1990)].

Table 2. Comparison of Horizontal Scales and Model Gird Sizes

A. Atmospheric process or feature	Typical horizontal scale (km)	B. Atmospheric model	Typical horizontal grid size (km)
Turbulent eddy	< 0.1	Urban-scale model	1–2
Power plant plume	1	Regional-scale model	10–100
Urban plume	50	Global-scale model	> 100
Convective cloud	10–100		
Synoptic weather feature	1,000		

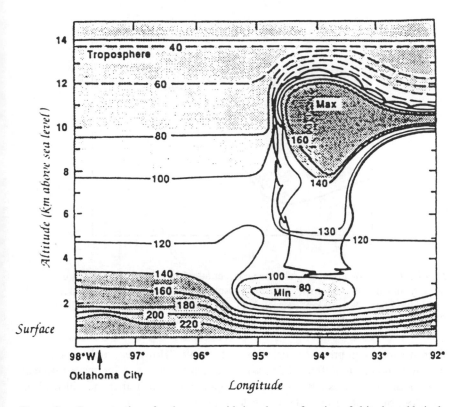

Figure 9. Concentration of carbon monoxide in ppbv as a function of altitude and latitude over Oklahoma City, OK., both outside and within a thunderstorm. [From Dickerson et al. (1987)].

The most common approach to dealing with important subgrid scale processes is to introduce into the model a parametrization that approximates the effects of the process based on the values of model-calculated variables; these parametrizations often are referred to as "subgrid scale parameterizations." As an example, most regional models now include some form of algorithm to identify regions of convective instability and then estimate the impact of these convective systems on the distribution and chemistry of the atmosphere using a parametrized description of cloud dynamics and microphysics. Similarly, Sillman et al. (1990) have demonstrated how the addition of parametrized plumes within each grid can be used to simulate more accurately the impact of both power plant and urban plumes on the photochemistry of the region.

A more sophisticated approach for addressing subgrid scale phenomena is through the use of "nested" grids. In this approach, a higher-resolution model with a finer grid structure is nested inside a coarser-grid model so that smaller-scale processes can be treated more comprehensively in some region of particular

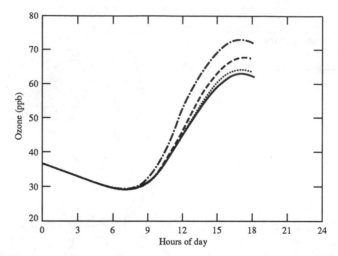

Figure 10. Regionally averaged ozone concentrations in ppbv versus time of day calculated by an eulerian model with horizontal resolution of 20×20 km^2 (solid line), 40×40 km^2 (dotted line), 80×80 km^2 (dashed line), and 400×400 km^2 (dot-dashed line). [From Sillman et al. (1990)].

interest such as an urban center (Rao et al., 1989). However, such an approach does not address the effects of subgrid processes outside of the nested domain and for this reason usually is used in conjunction with, rather than instead of, the adoption of subgrid scale parameterizations.

Evaluation and Verification of Regional-Air-Quality Models

Because of the myriad of complex processes treated in regional-air-quality models, it is imperative that these models be evaluated carefully before they are used in an operational sense, that is, in the design control strategies. This evaluation typically involves a comparison between observed chemical concentrations during a specific episode or set of episodes and model-derived concentrations obtained from a simulation or set of simulations designed to reproduce the observed episodes. Such a comparison between model-predicted and observed O$_3$ concentrations in the northeastern United States is presented in Figure 11. The agreement in the figure is good, and this result is typical of many similar evaluations that have been done for regional-air-quality models. (A specific test of a model designed to simulate ozone episodes in the southern United States has yet to be carried out, although calculations using EPA's ROM and NOAA's regional model are anticipated shortly as part of the Southern Oxidants Study.)

A major shortcoming of most evaluations of regional-air-quality models is the fact that they have been limited to a comparison of model-predicted and observed O$_3$ fields. The appalling paucity of high-quality observational data on the concen-

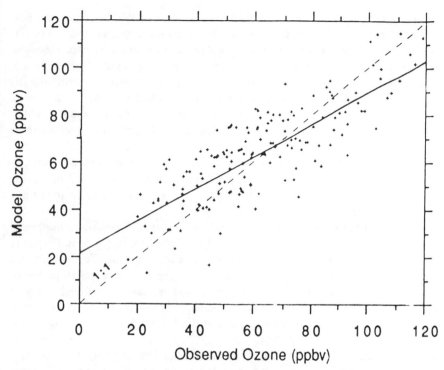

Figure 11. Comparison of model-predicted and observed ozone concentrations in the eastern and southern United States on July 6, 1986. Predicted concentrations obtained from the NOAA regional model of McKeen et al. 1991.

trations of VOC, NOx, and related species has precluded a more robust evaluation of regional-air-quality models. Thus, while the comparability in ozone concentrations is encouraging, it by no means represents a definitive evaluation of the accuracy of regional-air-quality models. Indeed, in light of recent studies which suggest that the VOC inventories used in regional-air-quality-model calculations underestimate the actual anthropogenic VOC emissions by a factor of 2 or more (NAS, 1991), the agreement between model-predicted and observed O_3 concentrations may be simply fortuitous and significant problems with the models may in fact exist. Until appropriate atmospheric datasets can be obtained so that a robust evaluation of the models can be done, the question of model accuracy will by necessity remain largely unanswered and the utility of these models to design future control strategies will remain limited.

Conclusion—Air Quality and Integrated Regional Models

The processes that control the formation of photochemical smog and its high concentrations of ozone in the northeastern and southern United States are myriad and complex. Sophisticated numerical models are essential for analyzing how these processes act in concert to bring about regional ozone episodes. For this reason regional-air-quality models represent the ultimate and most comprehensive summation of the physics, chemistry, and biology of regional air pollution, that is, the "state-of-the-science." However, because regional-air-quality models represent the cutting edge of the science, they should be viewed, at this time, more as research tools rather than as operational tools and large uncertainties are associated with their predictions. As detailed in the preceding section, these uncertainties are associated with a wide array of issues ranging from numerical techniques, to emission inventories, to the simulation of the relevant meteorological and photochemical processes. A major stumbling block toward improving these models is the present absence of suitable observational datasets to test and evaluate the models.

However, while the uncertainties in the algorithms used to describe atmospheric processes represent a major challenge for future research in regional-air-quality modeling, perhaps an even greater challenge is the development of new algorithms to treat the interactions among biological, sociological, and atmospheric processes that, while not presently treated in regional models, have a profound effect on regional air pollution and our ability to manage air quality over the long term. In the northeastern and southern regions of the United States where significant economic and technological changes are expected and where biogenic emissions represent a major source of VOC to the atmosphere, these interactions include: (1) the effect of growing population pressures, increased industrialization and urbanization and new technologies on land use and the concomitant impact of these land use changes on biogenic emissions of VOC from trees and other vegetations; (2) the potential for a positive or negative feedback between changing air quality in the region, the response of ecosystems to these air-quality changes, and the resulting increases or decreases in biogenic emissions from these responding ecosystems; (3) the effect of long-term climate change on biomes and energy usage in the region and their concomitant effect on air quality; and (4) the effect of changes in air quality on the economic development of the region (caused by impacts on agriculture and forests, as well as governmental regulations to control pollution) and the concomitant effect of these economic impacts on emissions and ultimately air quality. To study these interactions properly it will be necessary to develop a truly integrated regional model that treats not only the atmosphere but also the biosphere and society.

References

Adams, R. M., J. D. Glyer, S. L. Johnson, and B. A. McCarl. 1989. A reassessment of the economic effects of ozone on U.S. agriculture. *Journal of the Air Pollution Control Association* **39**:960–968.

Carter, W. P. L., F. M. Lurmann, R. Atkinson, and A. C. Lloyd. 1986. *Development and Testing of a Surrogate Species Chemical Reaction Mechanism.* EPA/600/3-86-031. U.S. Environmental Protection Agency, Research Triangle Park, NC.

Chameides, W. L., R. W. Lindsey, J. Richardson, and C. S. Kiang. 1988. The role of biogenic hydrocarbons in urban photochemical smog: Atlanta as a case study. *Science.* **95**:18569–18576.

Chameides, W. L., F. Fehsenfeld, M. O. Rodgers, C. Cardelino, J. Martinez, D. Parrish, W. Lonneman, D. R. Lawson, R. A. Rasmussen, P. Zimmerman, J. Greenberg, P. Middleton, and T. Wang. 1992. Ozone precursor relationships in the ancient atmosphere. *Journal of Geophysical Research* **97**:6037–6055.

Chang, J. S., R. A. Brost, I. S. A. Isaksen, S. Madronich, P. Middleton, W. R. Stockwell, and C. J. Walcek. 1987. A three-dimensional Eulerian acid deposition model: Physical concepts and formulation. *Journal of Geophysical Research* **92**:14681–14700.

Charmichael, G. R., and L. K. Peters. 1984. An eulerian transport/transformation/removal model for SO_2 and sulfate—I. Model development. *Atmos. Environ.* **18**:937–952.

Chen, W. Y., 1989. Estimate of dynamic predictability from NMC DERF Experiments. *Monthly Weather Review* **117**:1227–1236.

Derwent, R. G. 1989. A comparison of model photochemical ozone formation potential with observed regional ozone formation during a photochemical episode over the United Kingdom in April 1987. *Atmos. Environ.* **23**:1361–1371.

Dickerson, R. R., G. J. Huffman, W. T. Luke, L. J. Nunnermacker, K. E. Pickering, A. C. D. Leslie, C. G. Lindsey, W. G. N. Slinn, T. J. Kelly, P. H. Daum, A. C. Delaney, J. P. Greenberg, P. R. Zimmerman, J. F. Boatman, J. D. Ray, and D. H. Stedman. 1987. Thunderstorms: An important mechanism in the transport of air pollutants. *Science* **235**:460–465.

Dodge, M. C. 1989. A comparison of three photochemical oxidant mechanisms. *Journal of Geophysical Research* **94**:5121–5136.

EPA, 1991. *National Air Quality and Emissions Trends Report, 1990.* U.S. Environmental Protection Agency, Office of Air Quality Planning and Standards, Research Triangle Park, NC, EPA-450/4-91-023.

Gery, M. W., G. Z. Whitten, J. P. Killus, and M. C. Dodge. 1989. A photochemical kinetics mechanism for urban and regional scale computer modelling. *Journal of Geophysical Research* **94**:12925–12956.

Haagen-Smit, A. J. 1952. Chemistry and physiology of Los Angeles smog. *Indust. Eng. Chem.* **44**:1342–1346.

Heck, W. W., O. C. Taylor, R. Adams, G. Bingham, J. Miller, E. Preston, and L. Weinstein. 1982. Assessment of crop losses from ozone. *Journal of the Air Pollution Control Association* **32**:353–361.

Hov, O., E. Hesstvedt, and I. S. A. Isaksen. 1978. Long-range transport of tropospheric ozone. *Nature.* **273**:341–344.

IPCC, 1990. *Climate Change: The IPCC Scientific Assessment.* J. T. Houghton, G. J.

Jenkins, and J. J. Ephraums, eds. Intergovernmental Panel on Climate Change, Cambridge University Press, NY.

Isaksen, I. S. A., O. Hov, and E. Hesstvedt. 1978. Ozone generation over rural areas. *Environmental Science and Technology* **12**:1279–1284.

Korshover, J. 1976. *Climatology of Stagnating Anticyclones East of the Rocky Mountains, 1936–1975*. NOAA Technical Memorandum ERL ARL-55, Air Resources Laboratory, Landover, MD.

Lamb, B., A. Guenther, D. Gay, and H. Westberg. 1987. A national inventory of biogenic hydrocarbon emissions. *Atmos. Env.* **21**:1695–1705.

LeFohn, A. S., and J. E. Pinkerton. 1988. High resolution characterization of ozone data for sites located in forested areas of the United States. *Journal of the Air Pollution Control Association* **38**:1504–1511.

Lippmann, M. 1989. Health effects of ozone: A critical review. *Journal of the Air Pollution Control Association* **39**:672–695.

Logan, J. A. 1989. Ozone in rural areas of the United States. *Journal of Geophysical Research* **94**:8511–8532.

Lurmann, F. W., A. C. Lloyd, and R. Atkinson. 1986. A chemical mechanism for use in long-range transport/acid deposition computer modeling. *Journal of Geophysical Research* **91**:10905–10936.

McKeen, S. A., E.-Y. Hsie, M. Trainer, R. Tallamraju, and S. C. Liu. 1991. A regional model study of the ozone budget in the eastern United States. *Journal of Geophysical Research* **96**:10809–10846.

Middleton, P., J. S. Chang, J. C. Del Corral, H. Geiss, and J. M. Rosinski. 1988. Comparison of RADM and OSCAR precipitation chemistry data. *Atmos. Environ.* **22**:1195–1208.

NAS. 1991. *Rethinking the Ozone Problem in Urban and Regional Air Pollution.* Authored by J. H. Seinfeld, R. Atkinson, R. I. Berglund, W. L. Chameides, W. R. Cotton, K. I. Demerjian, J. C. Elston, F. Fehsenfeld, B. J. Finalyson-Pitts, R. C. Harriss, C. E. Kolb, Jr., P. J. Lioy, J. A. Logan, M. J. Prather, A. Russell, and B. Steigerwald. National Academy Press, Washington, DC.

Rao, S. T., G. Sistla, J. Y. Ku, K. Schere, and J. Godowitch. 1989. *Nested Grid Modelling Approach for Assessing Urban Ozone Air Quality*. Paper 89-42A.2. Presented at 82nd Annual Meeting and Exhibition of Air and Waste Management Association, Anaheim, CA, June 25–30.

Runeckles, V. C. 1992. Uptake of ozone by vegetation. In A. S. Lefohn (ed.). *Surface Level Ozone Exposures and Their Effects on Vegetation*. Lewis Publishers Inc., Chelsea, MA, pp. 157–188.

Samson, P. J., and B. Shi. 1988. *A Meterological Investigation of High Ozone Values in American Cities*. Report prepared for the United States Congress, Office of Technology Assessment. U.S. Government Printing Office, Washington, DC.

Schere, K., and E. Wayland. 1989. *EPA Regional Oxidant Model (ROM 2.0). Evaluation on 1980 NEROS Data Bases*. EPA-600/S3-89/057. U.S. Environmental Protection Agency, Research Triangle Park, NC.

Seinfeld, J. H. 1988. Ozone air quality models: A critical review. *J. Air Pollut. Control Assoc.* **38**:616–645.

Seinfeld, J. H. 1989. Urban air pollution: State of science. *Science.* **243**:745–753.

Sillman, S., J. A. Logan, and S. C. Wofsy. 1990. A regional scale model for ozone in the United States with subgrid representation of urban and power plant plumes. *Journal of Geophysical Research* **95**:5731–5748.

Stockwell, W. R., P. Middleton, and J. S. Chang. 1990. The RADM2 chemical mechanism for regional air quality modeling. *Journal of Geophysical Research* **95**:16343–16367.

Tesche, T. W. 1988. Accuracy of ozone air quality models. *Journal of Environmental Engineering* **114**:739–752.

Trainer, M., E. T. Williams, D. D. Parrish, M. P. Buhr, E. J. Allwine, H. H. Westberg, F. C. Fehsenfeld, and S. C. Liu. 1987. Models and observations of the impact of natural hydrocarbons in rural ozone. *Nature:* **329**:705–707.

van den Dool, H. M., and S. Saha. 1990. Frequency dependence in forecast skill. *Monthly Weather Review* **118**:128–137.

Venketram, A., P. K. Karamchandani, and P. K. Misra. 1988. Testing a comprehensive acid deposition model. *Atmos. Environ.* **22**:737–747.

Volz, A., and D. Kley. 1988. Evaluation of the Montsouris series of ozone measurements in the nineteenth century. *Nature* **332**:240–242.

Vukovich, F. M., J. Fishman, and E. V. Browell, 1985. The reservoir of ozone in the boundary layer of the eastern United States and its potential impact on the global tropospheric ozone budget. *Journal of Geophysical Research* **90**:5687–5698.

Zimmerman, P. R. 1979. *Determination of Emission Rates of Hydrocarbons from Indigenous Species of Vegetation in the Tampa/St. Petersburg, Florida Area.* EPA 904/9-77-028. U.S. Environmental Protection Agency, Atlanta, GA.

8

Integrated Models of Forested Regions

S. T. A. Pickett, I. C. Burke, V. H. Dale, J. R. Gosz,
R. G. Lee, S. W. Pacala, and M. Shachak

Introduction

This chapter presents the results of a discussion among physical, ecological and social scientists, to evaluate how and why forested regions should be the subject of integrated regional models (IRM). We present an overview of large landscapes that are actually or potentially forested as subjects for integrated regional models. By integrated, we mean models that deal with interactions among social, physical and ecological aspects of a system. We extend the scope of the topic somewhat from forests in the narrow sense to include areas where climatic change or human development can generate a woodland or savanna physiognomy. The interactions of climate, human actions and native ecological potential, which underlie such a gradient, suggest some of the utility of integrated regional models for understanding the structure and function of forested regions under changes in climate and land use.

This chapter has several specific goals: (1) to present reasons that forested landscapes are desirable, even critical, foci for integrated regional models, including practical and functional reasons; (2) to present illustrative questions that can motivate integrated regional modeling in forests; (3) to suggest in general terms the nature of the models that would integrate social, physical and ecological processes in forested landscapes; and (4) to present illustrative case studies, including metropolitan regions, arid regions, tropical forests and temperate forests slated for multiple uses.

Throughout the chapter, we indicate the existing excellent studies that have integrated ecological, social and physical aspects of forested landscapes. Because of the relative rarity of conceptual, simulation and analytical models that truly integrate the three areas of concern to us, much of our discussion is prospective and often hypothetical. We discuss a number of impediments to be overcome, and the need for a new emphasis to stimulate the growth of integrated modeling in forest regions.

Human–Environment Couplings Uniquely Integrate Regions

Forests as a Locus for Integration

Interactions among human, ecological and physical processes are critical to the structure and dynamics of forested regions. Many of the most pressing environmental and social issues of the day are occurring in forested or once-forested ecosystems (World Resources Institute, 1992). Roughly 500 million people depend directly on forests for survival (World Bank, 1991), deriving housing, fuel, and food from those forests. They, and the vast majority of the rest of the world's population, also depend on products obtained from once-forested landscapes. These uses include wood for structures, land for growing crops, pulp for paper, and plants whose products are used as medicines. The demand for forest products and arable lands results in a global pattern of cutting trees more rapidly than they can regrow (Williams, 1990; World Resources Institute, 1992).

In addition to their socioeconomic significance, forests are key to the ecological and physical functioning of the earth. Forests contain 73% of the organic carbon in the biosphere (Post et al., 1990), an estimated three times the amount in the atmosphere. Forests contain 90% of the carbon stored in terrestrial vegetation (Olson et al., 1985). Carbon sequestered in tree biomass has a relatively short turnover time (several centuries maximum), so that management strategies such as afforestation or deforestation can significantly alter the global carbon cycle over a human lifetime. Other ecosystems that hold significant stores of carbon, such as tundra or grassland, do not have turnover times that allow sequestration to occur at human time scales. Forested ecosystems play a key role in the hydrologic cycle, providing a large surface area for transpiration, and significantly influencing local and regional-scale weather patterns, runoff, erosion, and stream-flow (Shukla et al., 1990). Trees exchange a large number of other important gases with the atmosphere, including volatile organic compounds. In addition, forests provide the habitat for the majority of species living today on the Earth (Reid and Miller, 1989; World Resources Institute, 1992).

Finally, forests are particularly appropriate as the focus of integrated regional assessment because they are entities that are recognized by all disciplines and by the public. Boundaries of the forest biome are both defined by and define atmospheric interactions, such that climatic patterns change significantly outside of forested regions. Human patterns of use of the environment for resource production, jobs, and aesthetics are congruent with boundaries of forested systems. Thus, social scientists, ecologists, and atmospheric scientists have already developed their own conceptual and mathematical models for forested regions.

However, according to the well-corroborated theory proposed by Firey (1960), conservation or sustainability of forested regions is achievable only where the biologically possible, economically gainful and socially acceptable are integrated, and social and cultural conditions are stable enough to elicit voluntary conformity to nongainful as well as gainful resource practices.

Couplings Among Ecological, Social and Physical Factors

There are a number of interactions among human, ecological, and atmospheric systems that structure forested regions. Forest harvesting for sustained wood production and conversion of land for agricultural use are clearly the overwhelming controls over ecosystem structure, and the pattern of human disturbances over landscapes has major implications for plant community structure, wildlife habitat, organism migration, biodiversity, and exchange of energy and matter (Bierregard et al., 1992; Harris, 1984; Williams, 1991). Regional emissions of air pollutants such as sulfur dioxide, nitrous oxides, ozone, and hydrocarbons significantly influence the functioning of forest ecosystems, and likely limit the productive potential for resource use (Brown, 1993). Land use practices affect global emissions of greenhouse gases such as carbon dioxide, methane, and nitrous oxides that are increasing in their global atmospheric concentrations at unprecedented rates and are thought to be forcing climatic change. Such climatic change will result in altered geographic distribution of environmental constraints on forest distribution and forest growth (King et al., 1989; World Resources Institute, 1992). Human-mediated redistribution of species has resulted in a large number of environmental changes, including invasions of alien species and selective harvest of wildlife species that once played key roles in trophic structure.

Regional Couplings Are Understudied

The couplings between the processes studied by the natural sciences (e.g., ecology, hydrology, meteorology) and socioeconomic sciences are not well understood because of a lack of integrated, interdisciplinary approaches to problem solving. The tendency is for ecologists to model ecological systems and for socioeconomic scientists to model human systems or human values. At best, these separate activities are linked at a higher level than functional calibration. Extrapolating from separate biological and socioeconomic models in an "additive" fashion may not adequately represent systems behavior because interactions occur at levels that are not represented. In terms of hierarchy theory (Allen and Starr, 1982), each discipline has traditionally represented systems dynamics as a separate hierarchy of processes, and linkages among the hierarchies have only been considered at the highest levels. We suggest that lower-level, fine-scale interactions among disciplines or hierarchies must be considered in order to represent system dynamics adequately.

For example, human decisions about land use are directly related to socioeconomic constraints, but are also determined by opportunities provided by a landscape, e.g., potential productivity and timber yield. Explicit representation of these feedbacks is likely necessary for understanding current conditions and predicting changes (Fig. 1). Such disciplinary foci may result in a collective view that cannot see the forest (environmental interactions) for the trees (individual disciplines). Current ecological models, for example, would not remove beavers from the list of factors causing structure and functional change in a forest land-

LARGE SCALE; SLOW TURNOVER

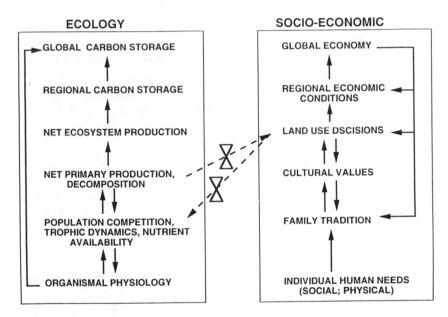

FINE SCALE; FAST TURNOVER

Figure 1. Representation of the traditional, separate, hierarchical conceptual models of the ecological and socioeconomic disciplines. Consideration only of upper-level interactions (e.g., between global climate and global economy) may provide statistical relationships, but cannot explain nor predict important feedbacks for future conditions. For example, land use decisions in a forested landscape control species composition directly, and via ecological interactions influence net primary production or potential timber harvest. The potential timber harvest directly controls future land use decisions (e.g., when to cut, how large an area, what species to plant, etc.). Through these interactions, land use decisions on a local scale may interact with the ecological hierarchy to result in regional carbon storage changes. Declining productivity due to system degradation in the ecological hierarchy may influence land use decisions and regional economic conditions. Consideration of either the disciplinary model alone or of only upper-level interactions would not consider these feedbacks or provide predictive power for understanding changing systems.

scape where that species occurred. Successful regional models should not *a priori* remove humans from the region. The hypothesis that humans inhabiting or using a landscape are an integral factor in that landscape must at least be tested. The growing awareness of the extent of human roles in ecosystems makes the hypothesis a compelling one (McDonnell and Pickett, 1993).

There are many impediments to the needed interdisciplinary integration, such as:

(a) Lack of a funding source. Direct funding for addressing these interdisciplinary problems would remove the necessity for researchers to participate as their time allows, e.g., only after completing other disciplinary tasks and research, or on a voluntary basis.

(b) Improper review process. Interdisciplinary proposals risk the wrath of two or more disciplines that prefer "mainstream" discipline science.

(c) A narrow academic reward system. Both reputation and tenure decisions tend to be made on the basis of publications within a discipline, sole- or first-authored publications, developing a presence in the "field," etc. Interdisciplinary science tends to be viewed as "weaker," nontheoretically based, or "merely" applied.

(d) Poor communication between the different disciplines (Dale, 1994). Differences in concepts and terminology, different approaches to problem recognition and problem solving, require significant amounts of time to resolve. This time is difficult to generate when other pressures resulting from within-discipline factors have high priority.

Although it is clear that the variety of constraints on integrated regional models will require much effort to overcome, perhaps funding is the key pressure point. If funding targeted on truly integrated regional models becomes more widely available, then the academic reward system, the commitment of investigators to improved interdisciplinary communication, and, perhaps slightly later, the appropriateness of the review process will certainly follow.

Questions Requiring IRM in Forested Regions

Within-Regional System Dynamics and Coupling

First, we must clarify our concept of "region." The concept of ecosystem entails a functionally connected suite of organisms and their physical environment. Interactions among organisms, flows of materials and energy, and changes through time have been foci for ecosystem studies. Although the fundamental ecosystem concept makes no assumption about homogeneity, the tradition has been to take ecosystems to be relatively homogeneous units (Kolasa and Pickett,

1991), often in order to simplify the study. When ecologists turned their attention to heterogeneity and to the role of spatial pattern in determining the functioning of ecological phenomena, the new discipline of landscape ecology was invented to house that interest (Forman and Godron, 1986; Turner 1989). Landscapes are considered to be large areas in which the composition of patches and the spatial patterns of patches (both often identifiable from maps) are controlled by geomorphology and disturbance. A region likewise is a collection of landscapes unified by climate, land use, and disturbance and is subcontinental in extent. Thus, the difference between landscapes, regions and continents is one of scale. Because the basic concern in both regional and landscape scales is with the influence of spatial pattern on process, and of fluxes that operate among patches, there is little to be gained from seeking a hard and fast criterion for differentiating between landscapes and regions. Indeed, the need for integrating the social, physical and ecological disciplines applies to both scales.

Many of the questions we present are not specific to forested regions; however, because forested areas are home to much of the world's human population, a large number of values or value systems applied to forests make them a good substrate for integrated regional models. The questions represent both ecological and social/economic viewpoints.

How does landscape structure and function within a region influence human activities, behaviors, and perspectives, and conversely, how do human influences affect natural processes (i.e., structure and function) within forested landscape? This question can be separated into a number of more detailed issues that focus on specific aspects of system dynamics. For example:

(a) How do natural and human-generated landscape patterns within a region affect natural disturbance regimes (e.g., fire, disease, insect infestations)?

(b) How does landscape configuration and change affect energy and matter flux, and influence atmospheric and aquatic pollution?

(c) How do landscape patterns affect organism movement (humans, native and exotic plants and animals), and the fluxes of commodities and economic processes? Do these processes feed back to influence subsequent landscape patterns?

(d) How do landscape patterns change through time as affected by the availability and spread of technology, economic changes, cultural changes, and climate change?

(e) How is forest use determined by regional activities (e.g., how is it related to fuel, building materials, pulp, agriculture, etc.) and how do past needs and historical or cultural practices influence current forest use?

(f) What are the socioeconomic controls on land use and how do these feed back to influence ecological processes?

   (i) What types of human activities result in immediate and dramatic changes in the natural structure and processes within forests and what forces (economic, demographic, legal, psychological, political, social) make humans take those actions?

   (ii) What types of human activities result in gradual changes in natural structure and processes and what forces motivate those actions?

   (iii) Who are the human "actors" that impact natural systems and respond to changing natural conditions?

   (iv) What is the balance between individual and collective decisionmaking? How do motivations for collective decisionmaking differ from those of individuals? What are the processes through which individuals and different types of groups make decisions?

   (v) How are values established? How are different types of values accommodated in decisionmaking? The specific issues that must be addressed in answering this question are the following:

     a. Reconciliation of short-term vs long-term criteria.

     b. Reconciliation of differences in opinion among people or groups regarding criteria.

     c. Reconciliation of pecuniary vs nonpecuniary criteria.

(g) How are regional patterns of carbon storage, biodiversity, or socioeconomic conditions affected by and affect land use patterns in the region?

(h) How do land tenure patterns or expectations influence land use pattern in regions? How are they influenced by environmental conditions?

(i) How do landscape patterns within the region affect quality of life?

Between-Region System Dynamics and Coupling

The aggregate pattern of regions and their individual properties may be linked to atmospheric sciences, socioeconomic sciences, or goals such as sustainability, in ways that are qualitatively different from within-region linkages.

Questions may focus on a particular region, but be concerned with the relationship of that particular region with other regions or other scales. Within-region focus involves questions such as:

1. How do human–environment interactions at the regional scale affect global climate change, and, in turn, how does climate change feed back into regional structure and function?

2. How do land use patterns affect carbon storage, trace gas emission, radiative balance, and water balance for larger areas such as continents?

3. How can integrated human–environment systems be optimally managed or developed so that they are sustainable (not all landscapes may be large enough for such optimization)? Of course, first we must determine the characteristics of an ecologically and socioeconomically sustainable system.

4. How do different kinds of diversities trade off in the optimization for sustainability?

5. Do sustainability of culture, resource base, and economy match?

In contrast to the concerns discussed above, in which a region is considered the target of influences from surrounding regions, the entire suite of regions can be examined as a whole. Such a between-regional focus involves questions such as:

1. How do regions interact with larger systems or with other regions? For example, how do land use management practices in tropical forests influence continental scale climate?

2. What are the implications of interregional interactions for land use patterns and human–environment interaction?

3. What fraction of economic, ecological, and physical control arise outside the region as opposed to within it? How much feedback is there between interregional influences and control within the region?

Questions Involving Feedbacks

Perhaps the most significant potential benefit of integrated regional models is their representation of dynamic, reciprocal feedbacks in ecological/social/economic systems (Fig. 1). This is qualitatively different from using the model output from one discipline as the input for a model from another discipline. Implied in all the specific questions below is the question of how best to model such feedback effectively.

1. Do IRMs provide an appropriate theoretical basis for analyzing multiple resource use strategies?

2. What are the relative magnitudes of anthropogenic and natural controls in a region, and how does the proportion shift with (a) scale; (b) spatial distribution of humans; (c) management strategy; (d) extraregional constraints?

3. What are the relative magnitudes of anthropogenic and nonhuman environmental controls on the socioeconomic component of the region? How

does the relative magnitude shift with (a) scale; (b) spatial distribution of humans; (c) management strategy; (d) extraregional constraints?

4. How do concepts developed in one of the source disciplines, e.g., sustainability in resource economics, change when they are employed in truly integrated regional models?

Questions Involving Scale

Because the processes that characterize the different disciplines that must contribute to integrated regional models may operate on different scales, the problem of how to link the various scales so as to expose the functional linkages between ecological, physical and social processes is fundamental. The following questions illustrate the concerns with scale:

1. At what scale does regional organization appear?
2. How do scales of organization change as human–environmental coupling changes?
3. Are there any scales at which the regional system simplifies (i.e., reduces to fewer or more direct interactions)? What models or technologies would be best suited to resolve such simplifications?
4. At what scales are goals such as sustainability, biodiversity, and human values optimized?

These questions are only illustrative. A different mixture of representatives from the three broad disciplines would undoubtedly generate a somewhat different list. However, the central issues addressed by these questions are almost certainly themes that will run through any set of questions about a spatial array composed of units that used contrasting currencies (money vs mass vs energy), operated on different temporal scales (e.g., an individual's lifetime vs a government's time in office), were based on different spatial scales (e.g., forest stands, aquifers, watershed), and were nested within other units. Indeed, all the questions we have raised deal in some way with the problems of differing currency or value, different spatial and temporal scale, nesting, and the transfer of influence or material among units whose identity is defined by one of the particular scales or phenomena. Reducing the problem to these abstract terms highlights and reinforces the need for integration.

Nature of Integrated Models

Forested ecosystems are uniquely poised for the development of integrated regional models. First, models already exist at the individual, community, landscape, regional and global scales. For example, there are many versions of calibrated and validated models of forest community and ecosystem dynamics (Shugart, 1984). Although development work continues, these models can currently predict responses of forests to anthropogenic management and disturbance (Dale and Doyle, 1987; Shugart, 1992), although such influences are currently often imposed as exogenous impacts. Second, because of the commercial value of forest products, there exists a wide variety of models of forest economics (Adams and Haynes, 1980; Mills and Kincaid, 1992). These are sufficiently mature and reliable to find routine use in economic decisionmaking. Third, recent models demonstrate that forested lands have significant interactions with the regional and global regulation of the atmosphere, including effects on regional precipitation and air pollution, and effects on the global regulation of atmospheric CO_2 (Houghton et al., 1983; Running and Coughlan, 1988; Shukla et al., 1990). These findings make inevitable the rapid development of atmosphere–forest ecosystem models. Fourth, sociologists, landscape ecologists, and geographers have developed a small number of models of the interplay among patterns of human land use in forested landscapes, ecosystem structure and function, and economics (Dale et al., 1993b; Parks 1992; Southworth et al., 1991). Although these fledgling integrated regional models are limited in scope (given the relative inaccessibility of funds to support this kind of work), they offer encouraging signs that the goal of integrated regional modeling is feasible. Some models have successfully predicted simultaneous changes in patterns of land use, economic activity, and ecosystem structure and function (Lee et al., 1992).

Thus, the necessary ingredients are in place. We have useful formulations of the components of integrated models of forested regions, some formulations that couple several components, and successful feasibility studies.

Model Development

Issues of scale are central to any modeling effort. Ideally, a model should accommodate the full range of spatial scales from the few square meters characterizing the effects of individual trees on ecosystem- and community-level processes, to the many thousands of square kilometers characterizing some atmospheric, economic and social processes (Fig. 1). Similarly, ecosystem–atmospheric feedback may occur at time-scales as short as a few hours (e.g., effects of transpiration rates on thunderstorm activity), and as long as several centuries (e.g., successful dynamics).

In practice, decisions about the scales of modeling are constrained by available data, mathematical and computational tractability, knowledge of relevant processes and the questions to be addressed by the model. Although there will obviously be a need to produce integrated regional models at a range of scales, we suspect that initial efforts will focus on two scales. First, we have a relatively large amount of data across the range of scales that characterize ecological field studies, stand dynamics, and fine-scale management decisions ("pixel sizes" of meters to hectares and days to one year). Second, remote imagery, atmospheric measurements and records of economic and social activity supply information at larger scales ("pixel sizes" from 10^3 m^2 to many thousands of square kilometers and one to ten years). The first class of data will lead to fine-grained models that might include existing forest stand and ecosystem models as components (e.g., FORET), while the second class will lead to coarse-grained models that might include the land use transition models developed by landscape ecologists (Gardner et al., 1993; Turner, 1989) or global vegetation models such as those using the Holdridge life zones (Emanuel et al., 1985).

A second fundamental impediment to the development of integrated regional models concerns the relationships among state variables of existing ecological, sociological, economic, and atmospheric models. For example, forest ecosystem models typically measure vegetation in units of grams of carbon, stand models in units of numbers and sizes of trees, socioeconomic models in units of value (economic, social or both), and physical models in units of atmospheric conditions (albedo, surface roughness and transpiration rate) or soil properties (nutrient content, particle size, etc.). Thus, modelers will have to define explicit mappings that translate one set of state variables into another. It is vital that several such mappings be investigated. For example, the relationship among forest structure, economic value and recreational value is at the heart of multiple land use planning.

Model Simplification

The success of any complex modeling effort depends not only on predictive power, but also on how much the model explains. Typically, explanatory power decreases as complexity increases (Levins, 1966).

A model is simplified by identifying schemes of aggregation (e.g. combinations of state variables averaged over some spatial and temporal scale) that produce new state variables governed by simple rules. When expressed in terms of the new aggregated state variables, the complex model becomes simple.

It is important to understand that the state variables of a simplified integrated regional model, and the temporal and spatial scales that characterize these variables, would be properties of the *integrated* system itself, rather than properties of any one ecological, socioeconomic or atmospheric component. Thus, the existence of appropriate rules of aggregation would establish the distinctness of the integrated field.

A second way to produce simple models is simply to *assume* an aggregation scheme and to formulate a simple model at the outset. The study of such models would serve to sensitize the imagination about the dynamics of integrated systems.

Case Studies

The utility, geographic spread, variety of kinds of forest and their interaction with humans and global processes call for the integrated regional approach to modeling and data collection under circumstances of afforestation, deforestation and continually forested areas. Therefore, four case studies are presented, which range across biomes, apply to contiguous forest or patches of trees, and are relevant to managed or natural forests. They span a range of environments and human situations from an arid area to a rainforest climate, and from sparsely populated to densely populated areas. We will present examples on forests in urban regions (McDonnell and Pickett, 1990), conversion of desert patches to savanna (Gillis, 1992), the Brazilian Amazon (Dale et al., 1993a), and temperate zone multiple-use forests (Franklin, 1993). These case studies demonstrate possible integrated models of currently, formerly, or potentially forested regions.

Forest in Urban Regions

Urban regions are an expanding land use type in the world (World Resources Institute, 1992). They are an amalgam of highly urbanized, suburban, exurban and rural land use types. In addition, the arrangement of land use types in metropolitan regions may generate a harlequin pattern on the land. Forests embedded in metropolitan areas are subject to a variety of interactions that can benefit from integrated regional modeling.

Objectives

• To study the role of forested areas in the context of an urban region.
• To provide scientifically sound input into the decision making process in order to establish or restore forested areas in metropolitan areas.

Question: How does the presence, size, configuration and function of forested areas interact with social and physical processes in metropolitan areas?

The scale of the model is set initially by the standard geographer's boundaries of a metropolitan region (Bourne and Simmons, 1982; Dickinson, 1966). Examples of focal variables include the following. Ecological variables are (1) plant and animal species diversity; (2) pollution loadings and effects; (3) forest regeneration; and (4) the role of urban water and nutrient flux. Social variables are (1) recreation; (2) transportation; (3) shopping and business; and (4) citizen image of and involvement with environment. Physical variables are (1) neighborhood microclimate; (2) water and air quality; (3) hydrological and flooding regime; and (4) interaction with anthropogenic pollutant plumes.

Background

A key issue for forests in metropolitan regions in practical terms includes questions about the amount of "green lungs" a city requires for air quality, temperature regulation and noise abatement. However, forests in cities, suburbs, exurbs and the nearby rural areas are also important for aesthetics, recreation, planning and biodiversity. Property values are, for instance, recognized to relate to the presence and quality of urban natural areas (Boyden and Millar, 1978). Because the role of forests in metropolitan areas affects and is affected by the social and physical environment, an integrated approach to such forests is required (Boyden, 1977; Stearns and Montag, 1974). The problem is of wide significance because the increasing proportion of human population that lives in cities and surrounding metropolitan regions (Frey, 1984; World Resources Institute, 1992).

Strategy

In the past, ecologists have largely ignored cities and have sought to study forests in non-urbanized areas. The boundaries between forests and their surroundings in metropolitan areas cannot be simply the obvious biological edge. Functions of such forests have conceptually "extended boundaries" that intimately involve human activities, institutional decisions and physical fluxes from the built component of the metropolitan area (McDonnell and Pickett, 1990). Therefore, full understanding of the role of forests in cities requires integration of ecological, social and physical processes (Boyden, 1977; Rapoport, 1993).

There are at least three steps to understanding urban forests (McDonnell and Pickett, 1990). First, basic forest data and landscape level pattern must be collected for metropolitan areas. This tactic allows framing initial working hypotheses about how forests are integrated with human and physical components of the metropolitan region. Second, it is important to study how the forest can act as a sink for pollutants or interact with biotic and abiotic fluxes in urban regions. Third, simple models can focus on the human–physical–forest interactions in the entire urban region as well as stratify by land use types.

Anticipated Results

The outcome of the strategy outlined above will be integrated regional models that can be used to improve planning the composition, location, size, management and control of flows of matter in urban forests. Environmental education will be enhanced by providing the public with information on local forested areas. Access to forested habitat will improve local science education in cities and suburbs. Restoration of damaged or impaired forests in metropolitan areas will be improved as a result.

Afforestation in Patches in Arid Land: The Case of Savannization

This case study focuses on an arid region in which the drainage patterns are being modified and trees are being planted. "Savannization" is the term given to this process (M. Sachs, personal communication) as practiced in the Negev Desert, Israel. It may well be that the larger region is one in which forests were much more important in the past. The process of savannization is appropriate in other parts of the world as well (Shachak et al., 1992).

Objectives

• To study changes in patterns and function in a desertified region under the influence of landscape modification that alters runoff flow (Yair and Shachak, 1987) and adds patches of trees.
• To measure and jointly model the effects of changes resulting from savannization on ecological, social, and atmospheric processes.

Question: How does the modification of landscapes to concentrate runoff and support patches of trees affect ecological, social and physical processes in the region, and the interaction among these processes?

The scope of the region is all areas of the Negev Desert between 100 to 300 mm annual rainfall (Evanari et al., 1982). Examples of focal variables include the following. Ecological variables are (1) plant and animal species diversity; (2) structural diversity of biota and habitat; (3) productivity; and (4) nutrient pools and fluxes. Social phenomena are (1) immigration from other regions and from other countries; (2) establishment of new settlements; (3) tourism from abroad; and (4) impact on the traditionally nomadic Bedouin culture (e.g., sustainability and integration with the national culture). Physical variables are (1) dry and wet deposition; (2) local microclimates; (3) albedo; (4) amount of rainfall; and (5) hydrological regimes.

Background

The issue of afforestation in desert lands is significant because of limitation of arable land in specific countries and throughout the world (Brown and Young, 1990; World Resources Institute, 1992). In Israel, absorption of immigrants from overseas and accommodation of a growing native population put population pressure on the Negev Desert. Worldwide, fully one-third of the land surface is classified as arid. Therefore, making arid lands available to support people without degrading recognized ecological values (e.g., biodiversity) is a critical need. In addition, the local traditionally nomadic population of Bedouin must continue to be supported in the desert. The region is important also as a tourist destination for Europe. Finally, the tactic of savannization can be used to counteract desertification, a widespread problem (World Resources Institute, 1992).

Strategy

In order successfully to understand the process and implications of savannization in the region, knowledge must be integrated from the start. Important background information on the ecology, climate, hydrology, geomorphology, social structure and attitudes, and political sciences are all required. The research group includes specialists in all the areas working together on a joint experimental design. All disciplines are included in the design from the beginning. Combination of research and development, stagewide and together, permit the integrated approach to have feedback from the development process, and to be immediately useful. The research is pursued on several time and spatial scales.

Anticipated Results

The project will provide both an empirical test and an integrated regional model that can guide further development in the region. Output will include information on the number and location of forest patches relative to ecological and social features of the region. Information will also be available on land use allocation in the region to enhance both biological and cultural diversity.

Tropical Deforestation in the Brazilian State of Rondonia: A Case Study of Land Use Issues

Objective

• To understand feedbacks between human patterns of migration and land use, the composition and structure of the forest ecosystem, and regional and global climate change.

Questions: What influences patterns of human migration into a region, and do these different influences result in different patterns of fragmentation of the forest ecosystem? How do different human land use systems cause different types of forest changes (e.g., loss of carbon, constraints or opportunities for dispersal or movement, effects upon biodiversity, soil degradation, changes in the frequency or intensity of natural disturbances, etc.)? How do changes in forest ecosystems produced by human uses affect the potential for future uses of the land? How will possible global climate and economic changes affect new human uses of tropical forest ecosystems (e.g., by changing the demand for agricultural products from temperate regions of the world)? What are appropriate strategies for ecologic and socioeconomic sustainability? What are the components of sustainability in the Amazon system?

Background

Movements of large numbers of people into tropical moist forests of the Amazon have resulted in massive deforestation and few socioeconomic gains for the people. Thus, dealing with deforestation combines socioeconomic and ecological components of land use change. Physical interactions enter in as a result of changes in the greenhouse gas emissions and subsequent effects upon global climate conditions and changes in moisture storage patterns affecting the regional climate, both of which feed back to affect socioeconomic decisions and environmental conditions. Significant changes in forest structures may induce microclimatic changes, which, in turn, may further alter patterns of vegetation, fauna, and human activities.

Large-scale human mitgration has a long standing and continuing effect on forests around the world (Repetto, 1988). Currently, for example, Guatemala and Indonesia are undergoing similar changes, and historically parts of the northern hemisphere forests have been cut in response to human population expansion (World Resources Institute, 1992).

Strategy

Multidisciplinary studies of tropical deforestation have documented the massive loss of forests, patterns of human movements, the inability of farmers to support themselves on the land, the trend toward pasture expansion (which has been supported by tax incentives), the degradation of agricultural and pasture lands, the release of carbon from the massive deforestation, and the effects of forest loss on the regional climate. Integrated studies are now being pursued that explore the effects of the carrying capacity of the forests on success with which humans can establish farms and ranches, the effects of controls on migration on deforestation, and feedbacks among particular land use strategy, forest loss and carbon emissions. The latter studies rely on computer simulations that model how social, economic and ecological factors interact to cause transitions in land uses of each site within a region and accumulate carbon emissions for the entire region (Southworth et al., 1991). A similar strategy can be used to project integrated effects upon biodiversity, soil degradation, and changes in the frequency or intensity of natural disturbances.

Anticipated Results

General results will be tools and concepts that integrate socioeconomic, ecological and atmospheric processes related to land use change. The framework for performing integrated studies will serve as an example for other studies (Dale et al., 1993a). A framework involves spatially explicit data on environmental conditions; social and economic surveys; a model that integrates social, ecological and atmospheric components; tabular and mapped output from the model; and statistical techniques to analyze temporal and spatial data. The development and testing of an integrated regional model that can simulate causes and effects of land use changes will be a major product. The model should be applicable to a variety of regions where questions of the effects of land use strategies are important.

The process of doing the research should result in better understanding of concepts that have a specialized meaning in particular fields but when interpreted across the integrated research program have broader meaning and applicability. An example of such an integrated concept is the idea of biodiversity, which can include the diversity of genes, species, communities, ecosystems, human ethnic groups, land use types, and habitats (including biotic structure and the edaphic and atmospheric components). Furthermore, new insights that are not possible from the single disciplinary viewpoint should result from the integrated approach. For example, we have already learned that the effects land use systems have on environmental conditions can restrict future land use opportunities and influence carbon dioxide emissions (Dale et al., 1993a).

Multiple-Use Forest Management

Objective

- To provide a mechanism for understanding and projecting the variety of land uses within a forest region. The research will necessitate explicit consideration of feedbacks among ecologic, socioeconomic and atmospheric processes.

Questions: How do land use planning and other management instruments affect patterns of land use in forested regions as well as physical and ecological processes (temperature and moisture gradients, biodiversity, disturbance regimes, movement of organisms, etc.)? How does forest fragmentation, changes in age and species composition, or soil degradation that results from human use constrain economic and social uses of the forest for future generations? In what ways may the predicted global climate change affect the integration of forest landscapes? For example, could warmer temperatures increase the demand for vacation homes in cooler forested regions of the more northerly areas or at higher elevations? How are decisions made to reconcile different values associated with forests? Such values range from the direct economic value of timber and other extracted products to aesthetic value, biological diversity and ecosystem equilibrium. Are there particular land uses that meet a number of objectives? Can multiple land uses within a region be arranged to best optimize objectives that arise from different disciplines? What tools are needed to address these multiple land use questions?

Background

The need for multiple land use management has been acknowledged for a long time. However, in the United States the conflicts between different land management activities has become even more pressing as forest resources are becoming scarce. The relative scarcity of particular forest types (e.g., old growth) has highlighted our need to understand the ecological interactions within those types. However, the regional approach forces scientists to resolve the unique attributes of forest systems that contain a variety of land use types, physical and environmental conditions, and socioeconomic pressures and histories. The potential for climate change to alter those forest systems also mandates the development of tools and approaches with which regional land management options can be explored. The development and use of integrated regional models will address basic research questions on the components and nature of feedbacks between socioeconomic, physical and ecological forces in managed forest regions (Dale and Gardner, 1987; Pastor and Post, 1988; Solomon, 1986).

Strategy

Models exist that combine the ecological and physical aspects of forest ecosystems. These models can project changes in forest composition, structure and function as a function of prevailing physical and environmental conditions and biological attributes of the component species. The forest ecosystem models can be applied to a region by running the model for each cell within a region or by stratified subsampling of the region.

Socioeconomic models of forest systems also exist. Forest economic models address questions of the affects of changes in available volume of commercial timber on regional economic growth. Regional population fluctuation can be modeled as a function of a variety of social and economic factors.

Output from socioeconomic models can drive forest models and vice versa, but currently there is no true integration of the models. However, there is a pressing need for integration so that the questions listed above can be addressed, and new concepts and links can expose novel questions. Integration could involve coupling of sets of models (with a focus on feedbacks between socioeconomic and ecological processes). On the other hand, key features of the existing models could form the basis for simpler integrated models.

Landscape models have incorporated certain socioeconomic and ecological aspects of forest management. These models produce transitions between land uses based on a combination of socioeconomic and ecosystem processes (Turner, 1987). Although the initial applications of these models have been relatively simple, the landscape approach holds promise for use as integrated regional models.

Anticipated Results

One of the primary results will be research frameworks for performing integrated regional studies that address management issues. These approaches should combine spatially explicit data and integrated models and concentrate on feedback between socioeconomic, ecological and physical processes. The new insights that arise from the integrated regional model approach will focus on interactions and will likely include how multiple-use management can serve a variety of objectives and the importance of spatial arrangement of management activities.

Summary

Forested regions provide a particularly important focus for integrated regional models because (1) they are intensively managed for human uses worldwide and (2) they provide a large and active terrestrial reservoir of carbon that interacts with the atmospheric pool. Hence, they have large impacts on global climate. We identify a set of questions that are appropriately addressed using models at a regional scale for forested systems, all of which will require integration of physical, ecological, and socioeconomic factors and disciplinary approaches. Several difficulties must be overcome in development of integrated forest regional models, such as the narrow scope of traditional funding and review processes, and the rigidity of traditional disciplinary boundaries. The forest biome is uniquely positioned for the development of integrated regional models because of the current status of linked atmosphere–ecosystem, vegetation–ecosystem, and socio-economic–ecosystem models that currently exist. We present four case studies that demonstrate the immediate need for such integrated regional models, including forests in metropolitan areas, intentional savannization of areas within a desert, deforestation in the Brazilian Amazon, and multiple-use forestry in temperate regions.

We suggest that integrated regional models of forested systems have a clear utility for assisting scientists in understanding large-scale interactions, and for planners, policy makers, and land managers in making wise decisions about land use. The understanding of the functioning, dynamics, and sustainability of forested regions will be advanced greatly through integrated regional models.

References

Adams, D. M., and R. W. Haynes. 1980. The 1980 timber assessment market model: structure, projections, and policy simulations. *Forest Science* **26**:1–64.

Allen, T. F. H., and T. B. Starr. 1982. *Hierarchy: Perspectives for Ecological Complexity*. University of Chicago Press, Chicago, IL.

Bierregard, R. O., T. E. Lovejoy, V. Kapos, A. A. dos Santos, and R. W. Hutchings. 1992. The biological dynamics of tropical rainforest fragments. *BioScience* **42**:859–866.

Bourne, L. S., and J. W. Simmons. 1982. Defining the area of interest: definition of the city, metropolitan areas and extended urban regions. *In* L. S. Bourne (ed.). *Internal Structure of the City*. Oxford University Press, New York, pp. 57–72.

Boyden, S. V. 1977. Integrated ecological studies of human settlements. *Impacts of Science on Society* **27**:159–169.

Boyden, S. V., and S. Millar. 1978. Human ecology and the quality of life. *Urban Ecology* **3**:263–287.

Brown, L. R. 1993. *State of the World: 1993*. W. W. Norton, New York.

Brown, L. R., and J. E. Young. 1990. Feeding the world in the nineties. *In* L. R. Brown (ed.). *State of the World: 1990*. W. W. Norton, New York, pp. 59–72.

Dale, V. H. 1994. Terrestrial CO_2 flux: the challenge of interdisciplinary research. *In* V. H. Dale (ed.). *Effects of Land Use Change on Atmospheric CO_2 Concentrations: Southeast Asia as a Case Study*. Springer-Verlag, New York, pp. 1–14.

Dale, V. H., and T. W. Doyle. 1987. The role of stand history in assessing forest impacts. *Environmental Management* **11**:351–357.

Dale, V. H., and R. H. Gardner. 1987. Assessing regional impacts of forest growth declines using a forest succession model. *Journal of Environmental Management* **24**:83–93.

Dale, V. H., R. V. O'Neill, and F. Southworth. 1993a. Simulating land use change in central Rondonia, Brazil. *Photogrammetric Engineering & Remote Sensing* **59**:997–1005.

Dale, V. H., F. Southworth, R. V. O'Neill, A. Rose, and R. Frohn. 1993b. Simulating spatial patterns of land-use change in central Rondonia, Brazil. *In* R. H. Gardner (ed.). *Some Mathematical Questions in Biology*. American Mathematical Society. Providence, RI, pp. 29–56.

Dickinson, R. E. 1966. The process of urbanization. *In* F. F. Darling and J. P. Milton (eds.). *Future Environments of North America*. Natural History Press, Garden City, NY, pp. 463–478.

Emanuel, W. R., H. H. Shugart, and M. P. Stevensen. 1985. Climatic change and the broad-scale distribution of terrestrial ecosystem complexes. *Climate Change* **7**:29–43.

Evanari, M., L. Shanan, and N. Tadmor. 1982. *The Negev: Challenge of a Desert*, 2nd ed. Harvard University Press, Cambridge, MA.

Firey, W. 1960. *Man, Mind and Land: Theory of Resource Use*. The Free Press, Glencoe, IL.

Forman, R. T. T., and M. Godron. 1986. *Landscape Ecology*. John Wiley & Sons, New York.

Franklin, J. F. 1993. Preserving biodiversity: species ecosystems, or landscapes. *Ecological Applications* **3**:202–205.

Frey, H. T. 1984. *Expansion of Urban Areas in the United States: 1960–1980*. USDA Economic Research Staff Report No. AGE5830615. Washington, D.C.

Gardner, R. H., A. W. King, and V. H. Dale. 1993. Interactions between forest harvesting, landscape heterogeneity, and species persistence. *In* D.C. Le Master and R. A. Sedjo, (eds.). *Modeling Sustainable Forest Ecosystems*. Washington, DC, November, 1992.

Gillis, A. M. 1992. Israeli researchers planning for global climate change on the local level. *BioScience* **42**:587–589.

Harris, L. D. 1984. *The Fragmented Forest: Island Biogeography Theory and the Preservation of Biodiversity*. University of Chicago Press, Chicago, IL.

Houghton, R. A., J. E. Hobbie, J. M. Melillo, B. Moore, B. J. Peterson, G. R. Shaver, and G. M. Woodwell. 1983. Changes in the carbon content of terrestrial biota and soils between 1860 and 1980: net release of CO_2 to the atmosphere. *Ecological Monographs* **53**:235–262.

King, A. W., R. V. O'Neill, and D. L. DeAngelis. 1989. Using ecosystem models to predict regional CO_2 exchange between the atmosphere and the terrestrial biosphere. *Global Biogeochemical Cycles* **3**:337–361.

Kolasa, J., and S. T. A. Pickett (ed.). 1991. *Ecological Heterogeneity*. Springer-Verlag, New York.

Lee, R. G., R. Flamm, M. G. Turner, C. Bledsoe, P. Chandler, C. DeFarrari, R. Gottfried, R. J. Naiman, N. Schumaker, and D. Wear. 1992. Integrating sustainable development and environmental vitality: a landscape ecology approach. *In* R. J. Naiman (ed.). *Watershed Management: Balancing Sustainability and Environmental Change*. Springer-Verlag, New York, pp. 499–521.

Levins, R. 1966. The strategy of model building in population biology. *American Scientist* **54**:421–431.

McDonnell, M. J., and S. T. A. Pickett. 1990. Ecosystem structure and function along urban-rural gradients: an unexploited opportunity for ecology. *Ecology* **71**:1232–1237.

———— (ed.). 1993. *Humans as Components of Ecosystems: The Ecology of Subtle Human Effects and Populated Areas*. Springer-Verlag, New York.

Mills, J. R., and J. C. Kincaid. 1992. *The Aggregate Timberland Assessment System Atlas: A Comprehensive Timber Projection Model*. General technical report PNW-GTR-281. USDA Forest Service, Pacific Northwest Research Station, Portland, OR.

Olson, J. S., R. M. Garrels, R. A. Berner, T. V. Armentano, M. I. Dyer, and D. H. Yaalon. 1985. The natural carbon cycle. *In* J. R. Trabalka (ed.). *Atmospheric Carbon Dioxide and the Global Carbon Cycle*. Volume DOE/ER-0239. U.S. Department of Energy, Washington, DC, pp. 175–214.

Parks, P. J. 1992. Models of forested and agricultural landscapes: Integrating economics. *In* M. G. Turner and R. H. Gardner (eds.). *Quantitative Methods in Landscape Ecology*. Springer-Verlag, New York, pp. 309–322.

Pastor, J., and W. M. Post. 1988. Response of northern forests to CO_2 induced climate change. *Nature* **334**:55–58.

Post, W. M., T.-H. Peng, W. Emanuel, A. W. King, V. H. Dale, and D. L. DeAngelis. 1990. The global carbon cycle. *American Scientist* **78**:310–326.

Rapoport, E. H. 1993. The process of plant colonization in small settlements and large cities. *In* M. J. McDonnell and S. T. A. Pickett (ed.). *Humans as Components of Ecosystems: The Ecology of Subtle Human Effects and Populated Areas*. Springer-Verlag, New York, pp. 190–207.

Reid, W. V., and K. R. Miller. 1989. *Keeping Options Alive: The Scientific Basic for Conserving Biodiversity*. World Resources Institute, Washington, DC.

Repetto, R. 1988. *The Forest for the Trees? Government Policies and the Misuse of Forest Resources*. World Resources Institute, Washington, DC.

Running, S. W., and J. C. Coughlan. 1988. A general model of forest ecosystem processes for regional applications. I. Hydrologic balances, canopy gas exchange, and primary production processes. *Ecological Modelling* **42**:125–154.

Shachak, M., B. Boeken, J. Cepeda-Pizarro, J. Gutierrez-Camus, J. Wrann, S. Benedetti, W. Canto, and G. Soto. 1992. *Savannization, an Ecological Answer to Desertification: A Proposal for a Savannization Project in Chile*. Proceedings of the Savannization Workshop at the Universidad de la Sirena, Chile, November 1992.

Shugart, H. H. 1984. *A Theory of Forest Dynamics: The Ecological Implications of Forest Succession Models*. Springer-Verlag, New York.

Shugart, H. H. 1992. Global models of change based on species and/or functional groups. *Annual Review of Ecology and Systematics* **23**:15–38.

Shukla, J., C. Nobre, and P. Sellers. 1990. Amazon deforestation and climate change. *Science* **247**:1322–1325.

Solomon, A. 1986. Transient response of forests to CO_2-induced climate change: simulation modeling experiments in eastern North America. *Oecologia* **68**:567–579.

Southworth, F., V. H. Dale, and R. V. O'Neill. 1991. Contrasting patterns of land use in Rondonia, Brazil: simulating the effects on carbon release. *International Social Sciences Journal* **130**:681–698.

Stearns, F., and T. Montag. 1974. *The Urban Ecosystem: A Holistic Approach*. Dowden, Hutchinson, and Ross, Stroudsburg, PA.

Turner, M. G. 1987. Spatial simulation of landscape changes in Georgia: a comparison of three transition models. *Landscape Ecology* **1**:29–36.

———. 1989. Landscape ecology: the effect of pattern on process. *Annual Review of Ecology and Systematics* **20**:171–197.

Williams, M. 1990. Agricultural impacts in temperate lands. *In* M. Williams (ed.). *Wetlands: A Threatened Landscape*. Blackwell, Oxford, pp. 181–206.

World Bank. 1991. *The Forest Sector*. The World Bank, Washington, DC.

World Resources Institute. 1992. *World Resources 1992–1993*. Oxford University Press, New York.

Yair, A., and M. Shachak. 1987. Studies in watershed ecology of an arid area. *In* L. Berkofsky and M. G. Wurtele (eds.). *Progress in Desert Research*. Rowman and Littlefield, Totawa, NJ, pp. 145–193.

IV

Summary and Overview

9

Prospects for the Development of Integrated Regional Models

Elizabeth Blood

The primary motivation for this book was the perception that the scale of human-accelerated environmental change necessitates conceptual and mathematical models of regional scope. Current modeling of human and environmental change has been discipline specific (physical, ecological, social) and generally focused at the individual, ecosystem or global scale. Models to integrate physical, ecological and social systems will be necessary to solve critical environmental problems. Currently, it is not clear how to facilitate such an integrated activity, but it was perceived that an initial step was to integrate the different disciplinary communities with their characteristically different focus on scales and processes and divergent vocabularies. The purpose of the workshop was to bring together people from different disciplines to inform scientists about the current state of knowledge about processes operating at a variety of scales, including regions, explore mechanisms by which scientists from different disciplines can work together to advance their collective knowledge of critical processes, and to improve both mathematical and conceptual models within and bridging their disciplines, identify steps necessary to move toward the development of integrated regional models (IRMs) that represent linkages across ecological, human and physical systems and identify the data requirements necessary to do successful integrated regional modeling. A major outcome of successful integrated regional modeling would be improved decisionmaking and management applications.

Integrated regional modeling will require the development of disciplinary models that can be linked through flows of relevant and important commodities (Fig. 1). New concepts of structure, function and change will need to be developed which transcend current modeling efforts. New "simple structures" will be necessary that can robustly embody the complexity of the subsystems composing them. As an example, one could develop a response surface diagram that embodies the multiscale relationship between a system process and its temporal or spatial resolution. Salt marsh cordgrass primary production is dependent on tidal inunda-

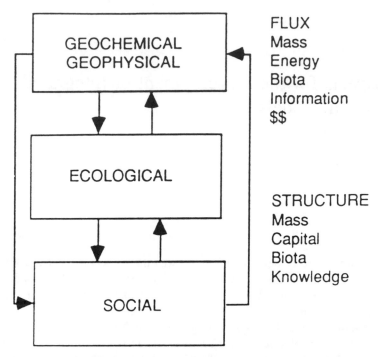

Figure 1. A conceptualization of integrated regional modeling as a function of specific discipline.

tion, but that relationship varies with temporal scale. The relationship between daily to decadal inundation variation could be incorporated into a series of temporal functions (see Fig. 2). On a daily to monthly basis tidal inundation duration and extent would have a positive effect on primary production by increasing the flushing of reduction products, but on annual to decadal scales would result in a decrease due to increased anaerobic conditions and a build up of reduction products. In crossing between disciplines, common currencies or flows could be functionally incorporated. An example would be the relationship between production and consumption in human and natural systems. Are the functional relationships the same and do they vary in a similar manner with "succession" or "evolution" of the connected systems? These simple structures should portray flows within and between the disciplinary models that encompass both qualitatively and quantitatively rich information that can be decomposed into disciplinary specific relevant information.

The challenges of doing integrated regional scale modeling fall into three broad categories: (1) creatively dealing with issues (calibration, validation, uncertainty and error propagation, simplification or aggregation, resolution and scale) inherent in using modeling as a mode of operation, (2) the state and availability

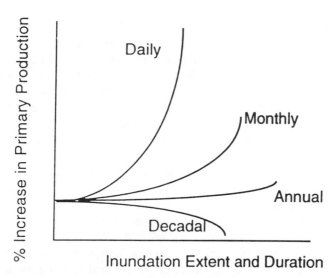

Figure 2. Represents the functional relationship between tidal inundation extent and duration and salt marsh cordgrass primary production increase and the scale of increased inundation. If the inundation is greater per day, it would enhance "flushing" of reduced sediment solution products (e.g., sulfides), but if the inundation is increased on annual to decadal scales, the daily variation would decrease and result in a build up of reduced products and, therefore, decrease potential production. A number of more complex relationships of hydrology, soil solution chemistry and physiology are embedded within this "simple structure" relationship of a control and system response.

of knowledge in the individual disciplines, and (3) the most challenging will be in interfacing the disciplines into a integrated model. Developing integrated regional models will require a melding of social, ecological and geophysical disciplines to provide sufficient information necessary to capture the integrated overall dynamics and processes occurring at the regional scale. Within each discipline it will be necessary creatively to synthesize inherent complexity into simple, yet robust structures of form and function. These structures must capture the arrangement and interrelationships among important components of the individual levels. Although the individual disciplines can provide detailed conceptualization, the real challenge will come in developing simple structures that embody the detail within each discipline. The frontiers of IRM are at the interfaces between the disciplines. The breakthroughs will occur in developing the flows and linkages between levels (disciplines) or finding a central paradigm that can integrate the disciplines. Developing a "common currency" that can flow between the levels (i.e., disciplines) will require significant interaction among the scientists and managers.

Integrated regional modeling forces explicitness and synthesis. It demonstrates the importance of interdependence of one level on other levels. With IRMs scientists and managers can explore alternative solutions to complex environmental problems (particularly when human systems are involved). The approach forces multidimensional understanding of issues, policy and science. These models would explicitly incorporate the "reality" (humans) of current ecological systems. Such modeling would promote or foster intellectual leaps within disciplines and new breakthroughs in knowledge at interfaces between disciplines and may not be possible at smaller integrating scales.

Perhaps the most general and compelling argument for the need to integrate regional models is that interactions among human, ecological, and atmospheric processes are critical to the structure, dynamics, and vitality of regions. And simply, it is not known how to integrate effectively these models at the regional scale. This gap in our scientific abilities comes at a time when human–environmental interactions are causing the greatest rate of change the Earth has seen in over 10,000 years.

Attendees at the workshop agreed that interactions among human, ecological and atmospheric systems largely occur at the regional scale (areas larger than ecosystems or landscapes but subcontential) and that these interactions drive regional structure. For example, environmental policy, economic incentives and return rates are governed by political units that largely correspond to the regional scale. Cultural and social traditions that control the ethics of land use tend to occur locally but are confined at regional scales. Atmospheric scientists have suggested that air-parcel dynamics and interactions with characteristics of the Earth's surface that occur at the "mesoscale" (e.g., region) are key processes determining regional climate patterns. Collectively, interactions among human, ecological, and physical forces both define and structure regions, and form the basis for interdisciplinary study at this scale.

A second general conclusion is that the couplings of natural (e.g., environmental) and human (e.g., socioeconomic) sciences are not well understood. There have been few integrated, interdisciplinary approaches to problem solving. The tendency is for ecologists to model ecological systems and for socioeconomic sciences to model human systems or human values. At best, these separate activities are linked at some later activity rather than initially building an integrated model as part of a focused effort. Extrapolating from two models in an "additive" fashion will not reproduce the results from a truly integrated model of all relevant factors. For example, current ecological models would not exclude the activities of large mammals in shaping ecosystem structure. Successful regional models should not exclude human activities.

A major factor that arises from the inclusion of humans in IRM is concerns over "values." There was quite a large difference in the view of the role of humans within an IRM. These perceptions were related to human values and how they could be measured and quantified (perceptions, quality of life, human health, willingness to pay, distribution, equitability). For many topics there will be questions about the "Societal Value" of an IRM effort and concern about policy implications and societal misuse of the model. Such concerns will lead to an increased sensitivity to "accountability" due to the imposition of societal values. This concern about values is an important departure from the historical perspective of ecological models as being value neutral.

Another major departure from current ecological models that arises from the inclusion of humans is a shift in emphasis from "advancement of the science" to more of an applied focus. Better management of resources, providing a means to play out management alternatives and incorporation of policy issues have been raised as justifications for IRM development. Such an applied focus is a shift from the usual justification of modeling. Modeling has been viewed as a synthesis mechanism for science inquiry, a predictive tool to extrapolate principles or concepts into the future, and a means to identify weaknesses in information or knowledge. The historic emphasis has been on science advancement, not on providing a tool to solve societal problems.

In all disciplines, modeling serves as a mode of integration and as a tool for identifying inherent uncertainties and feedbacks among components that may complicate the understanding and tractability of a problem. One of the major values of modeling is in identifying unexpected factors and interactions that prove to be important and providing insights to weaknesses in assumptions and hypotheses. Several factors that modeling efforts in different disciplines have in common were brought out in the previous chapters including:

1. Scale and feedbacks across scales (I. Burke).
2. Resolution—spatial and temporal (W. Chameides).
3. Uncertainty and error estimation (R. Berk).
4. Validation—evaluation (R. Berk).
5. Humans—how to explicitly incorporate differing views of social scientists (D. Liverman).
6. Aggregation—balance between simplification and complexity while embodying sufficient relationship to the original problem. The art is in the judgment of the importance of inclusion or exclusion of a given element to the essential functionality of the model versus its contribution toward unmanageable complexity. Regional systems are inherently complex.
7. Bounding.
8. Dichotomy between individual and systems approach.

The workshop participants believed that the outcome of a successful IRM activity would be a substantial advancement in the collective ability of scientists to adapt to a rapidly changing world. There would be a better appreciation of other disciplinary approaches and concerns that are contributing to regional issues. There would be an opening of dialogue among groups that previously were largely isolated. This interaction would result in the development of active cooperation among the disciplines and stimulate thinking far beyond current boundaries. As a result, it would create unusual opportunities for the cross fertilization of disciplinary ideas and concepts.

To stimulate the development of IRM, a number of fundamental needs must be considered. These needs relate to the availability of information (e.g., data) in a usable form, availability of knowledge and expertise, the ability to develop a common language for effective communication (e.g., overcoming jargon), and the identification of a common currency for model validation. Finally, there is a real need to develop several successful activities (case studies) for the broader community to learn from and to build upon.

Despite the generally enthusiastic support for an IRM activity there are several concerns to be considered. These concerns are:

1. That IRMs are not a panacea for all issues. Administrators and funding agencies should be careful not to force IRM activities beyond appropriate limits or force them to examine inappropriate issues.

2. IRM activities may encourage a strong "top-down" administrative structure. This approach is not perceived by the scientific community as needed or appropriate. Any "top-down" administration must be coordinated carefully with "bottom-up" initiatives.

3. The scientific risk for individuals within disciplines is real. Academic and institutional departments may not recognize the value of interdisciplinary work especially if done by younger professionals or by scientists with a tradition of departing from the mainstream of the disciplinary science.

4. The source of funds used for this activity will be suspect if the monies are reallocated from existing programs. The IRM activity should be funded from new monies even though it is recognized that some existing funds will need to be used and the importance of the activity is of overwhelming importance. New monies will help build broad-based support for the IRM activity whereas reallocation of funds will not.

Even with these basic concerns the participants strongly endorsed the concept of an IRM activity. The specific recommendations are as follows:

1. Integrated Regional Models should be developed to provide the basis for understanding the multiple interactions of people with the geophysical/geochemical and ecological process at the regional scale.

2. The IRMs should explicitly seek to provide a realistic framework for assessment of environmental and resource issues and the impacts of proposed policy and management regimes.

3. Each of the participating modeling communities should be pushed to

articulate the information that each requires from the other in order to address the significant interactions among and within their respective modules. Workshops bringing together the three modeling communities could accelerate this process. These workshops could also provide an opportunity to develop more precise ideas about the information flows that will connect the modules within IRMs.

4. As a first step there should be a critical evaluation of existing regional models. There is a real need to learn from both the successes and failures of existing efforts. As part of the process of developing IRMs, means should be developed to assess and evaluate uncertainties associated with input data and model structure from the diverse disciplines that contribute to IRMs.

5. The initial focus should be on a scale where socioeconomic, ecologic, and geophysical models overlap to address real societal issues. These models should be developed for a variety of chronological (historic) and geographical situations rather than on a continuous historical basis. This recommendation recognizes the difficulty of modeling social and ecological discontinuities.

6. The initial efforts should utilize simple, but robust models, that focus on the interfaces between human and environmental sciences. These models will need to be aware of common disciplinary concerns and issues as well as be of scientific and societal value.

7. The development of an educational training program will be essential for the long-term success of an IRM activity. The educational program will need to include pre-university students, traditional undergraduate and graduate education, as well as post-university professional education.

8. Although several federal agencies and departments will benefit greatly from this activity (e.g., EPA, NASA, NIH, Interior), it is strongly recommended that the National Science Foundation take the responsibility for obtaining and integrating funding and for program development. The exploratory nature of the complex task at hand and the traditional role that NSF has played in basic scientific research are key reasons for this recommendation.

9. It is recommended that strong interdisciplinary panels be established to review proposed research and requests for funding. Nitpicking by disciplinary experts, although potentially valuable to the specific pursuit, will not foster an environment to develop IRMs. Interdisciplinary panels hopefully can evaluate detail within the larger goals.

10. Although appreciable data have been collected and are available, there still exists a need to improve access to diverse data and to balance

the collection of data to develop functional and applicable IRMs. For example, data are available on the supply of fuels, but very little on how fuels are used. Consequently, there is relatively little information on energy efficiency and how it has changed over time in different economic sectors. Modeling can be helpful in guiding data collection in such cases.

Rambo (1983) used human ecology to develop a general systems theory that might prove useful in integrating regional modeling. Human ecology is the interaction of two open systems (social and ecological), which receive inputs and outputs from one another. These systems exist in "a complex, dynamic relationship with multicausal, multidirectional exchanges of energy, material, and information. Inflows are incorporated into form, function and inter-relationships within the system through processes of consumption, production and information exchange. Each system is open to external influences through diffusion, migration, and colonization" (Turner et al., 1990). Changes in the system may be sudden or gradual and adaptive, with evolution expressed as survival of species and choices of individuals institutionalized as social norms.

Human ecology is an approach that could bridge the gaps between natural and social sciences. This discipline incorporates nature's constraints on human behavior and human behavior feedback on the environment. The approach focuses on connectivity and mutual causality among natural and human components. This approach embeds humans in the natural environment and allows us to understand what we are doing and why. Natural sciences are focused on the quantification of energy, materials, and information with an emphasis on interconnections, dynamics and exchanges with the external environment. Organisms are linked through the exchange of energy according to the laws of thermodynamics and materials through biogeochemical cycles. System equilibrium is not a fixed point but a temporally varying equilibrium distribution. A dynamic nonlinear, quasiequilibrium is stressed with changes occurring through evolution and succession.

Human ecology emphasizes social organization, human biology, technology and "values." Social relations are materialist and focus on the transfer of goods, services and energy. Through the same flows of energy, materials and information that order ecological systems, social systems are ordered through modes of production to produce social and economic formations. This approach incorporates the forces of production (e.g., technology, labor) and the relations of production (e.g., human–human interaction). As an example, availability of clean water requires that human wastes be treated before release into surface waters. Water treatment is dependent on technology and labor. Waste water treatment results from humans interacting through social systems (taxes, regulatory institutions, laws). Humans must place a "value" on clean water, develop a plan, and set social goals before the social system can function to provide the structures necessary for water treatment. By using modes of production, human institutions and forms of human "consciousness" can be incorporated into a multitiered, dynamic model of human systems.

The dichotomies between the theories of nature and social science are brought together through "consumption." Consumption brings together the realms of meaning, nature and social relations and in the process transforms each. Through consumption, we withdraw resources and produce residuals and secondarily establish technical and physical infrastructures, thereby interconnecting with and transforming nature. A product embodies social and economic relations through influencing meaning (e.g., through advertising) and social status through context, place and wages. This in turn affects human perceptions about nature.

References

Rambo, A. T. 1983. *Conceptual Approaches to Human Ecology.* Research Report No. 14. Honoluli, HI: East-West Center.

Turner, B. L., W. C. Clark, R. W. Kates, J. F. Richards, J. T. Mathews, and W. B. Meyer. 1990. *The Earth as Transformed by Human Action.* Cambridge University Press, New York, NY.

Index